S0-COM-655

Computing in
Clinical Laboratories

COMPUTING IN CLINICAL LABORATORIES

Edited by

Frederick Siemaszko BSc

Principal Scientific Officer
Department of Chemical Pathology
Bristol Royal Infirmary

A Wiley Medical Publication

JOHN WILEY & SONS

New York/Toronto

John Wiley and Sons Inc.
605 Third Avenue, New York, NY 10016

First published 1978 by
Pitman Medical Publishing Co Ltd
P O Box 7, Tunbridge Wells, Kent, England

ISBN: 0-471-04321-4
Library of Congress Catalog Card Number:
77-94061

Printed and bound in England
by The Pitman Press, Bath, Avon

CONTENTS

Part 7

Recent Advances in Numeric Techniques

FOREWORD

Professor T.P. Whitehead

The Second International Conference on Computing in Clinical Laboratories held in Birmingham, England attracted over 300 participants, over twice the number that attended the First Conference. Participants came from 16 different countries and included computer scientists working in industry and clinical laboratories as well as clinical laboratory scientists representing the various disciplines associated with clinical laboratory investigations.

Frequently there are four phases in the application of technological innovations in medicine. The over-optimistic introductory period lead by enthusiastics, a second phase of developing pessimism as the limitations and full costs of the innovation are realised. The third phase is a recovery phase where a balancing of the claims of phase one and two is achieved and finally the fourth phase where the innovation is applied at its correct level.

The introduction of computers to clinical laboratories is no exception to this general phenomenon and for many aspects of the subject we are somewhere between stages three and four.

The papers describing the use of computers in the selection, use and interpretation of laboratory results are illustrative of a maturing assessment of the computer's role.

We are still at an early phase in assessing cost effectiveness. There are reasons for this and these are emphasised in this book; technological innovation continues to make even recent cost calculations out-of-date. The introduction of microprocessors as integral parts of clinical laboratory apparatus is an example of such innovation. It is gratifying to see that the difficult problem of patient and specimen identity appears to be being resolved by the use of bar coding.

It is difficult for me to envisage working as a clinical chemist without a computer in the laboratory. For this reason I enjoyed the Birmingham Conference and I am sure that those who were prevented from attending will enjoy reading the presented papers; they indicate several milestones of achievements and pointers to the future.

Part 1

SYSTEM DESIGN AND IMPLEMENTATION

PRINCIPLES OF LABORATORY SYSTEM DESIGN

P.J. Vitek
Division of Computing and Statistics,
MRC Clinical Research Centre,
Harrow, Middlesex HA1 3UJ

Professor M.J.R. Healy
Department of Medical Statistics and Epidemiology
London School of Hygiene and Tropical Medicine
Keppel Street, London WC1E 7HT

INTRODUCTION

Many pathology laboratories have now installed computer systems and a good deal of experience with them has been gained. The large majority have been in clinical chemistry reflecting the growth of demand for tests and the progress in automation in this discipline, but systems for haematology, microbiology and histopathology also exist.

It is by no means an easy task for a laboratory director to decide whether to install a computer system, and if so, what kind it should be. The range of tasks that such a system might undertake is wide, the possible methods for each task are numerous, and above all the cost, not only of the capital investment but also of operating the system, may vary substantially. Moreover, once the choice is made and implemented, it may be difficult to change it, particularly if the desired modification was not envisaged when the system was planned. The object of this paper is to set out some of the possible considerations on which the system design should be based together with some of their implications.

SYSTEM DESIGN

An essential first stage in designing a laboratory system is to undertake a thorough system study of laboratory operation, with a view to defining the genuine problems and delineating areas in which computer methods would provide useful solutions. This should not, however, be limited to describing existing laboratory procedures but an effort should be made to devise more rational ones. The proposal for installation of a computer system must specify what tasks it is to perform and in particular how its installation is to affect the manner in which the laboratory is to run.

The result of such a study should be a selection of basic functions, a 'backbone' of the system, which will then support and control other functions or applications. Separation of basic functions from applications is a fundamental concept in system design. An application may be changed, deleted or added without disturbing the system operation (e.g. data collection from a new analyser or the introduction of ward lists), but to change some basic, organisational function of the system implies not only a major re-programming task but also a substantial change in laboratory operation; as an example, we could consider splitting the work flow through the laboratory into several independent streams with separate requests.

The hardware requirements of the computer system can vary over a wide range according to the tasks to be performed; sensible decisions upon the type and the amount of equipment and the accompanying basic operating system cannot be reached until these tasks have been specified in some detail. There is plenty of scope for choice, both in what should be done and in how it could be done.

What needs emphasis is that installing all but the simplest of computer systems will cost about as much as a multiple analyser and will demand at least as thorough a rethinking of all laboratory practices as normally precedes the installation of one of these devices. The prospect of machine breakdown must be squarely faced in each case. Just as manual methods of analysis or single channel instruments must be able to stand in for a multiple analyser for long enough for it to be repaired, so a computer system must not be commissioned without detailed plans for falling back on manual techniques.

Another important part of the system design is specification of interfaces through which the laboratory system will communicate with the users and other systems, existing or future ones. The quality and quantity of the information received by the laboratory will, to a large extent, determine the standards of performance of the system. Specification of the system interface must also include allocation of responsibilities for its maintenance; it is not sufficient to say what the system should receive or produce but also who will ensure that the necessary information is available.

It is now beginning to be recognised that acceptance of the computer system by the laboratory staff plays an important part in its success or failure. Existence of a badly designed 'human' interface is usually manifested by extra work for technicians and clerical staff (justified simply by saying 'it must be done for the computer') or by a loss of flexibility in performing the analytical and clerical procedures. Special attention should be given to flexibility of the system not only because of rapid technological progress in developing new laboratory equipment but also because of the not infrequent need to implement new tests or analytical techniques.

Finally, the system specification, which should be the result of the system design, is a prerequisite for the successful design of programs and their implementation. The importance of a correct system design is highlighted by the recent developments in semimechanical techniques for production of programs, e.g. Jackson(1). These techniques ensure that the programs will be 'correct' in the sense that they will perform the specified functions. It is, however, beyond the powers of a programmer to salvage an incorrectly designed system.

THE SCOPE OF LABORATORY COMPUTING

It is not easy to decide what tasks the laboratory computer system should do. The outside computer specialist entering a hospital laboratory for the first time is liable to have his attention focused on the array of mysterious analytical equipment, and it is not altogether surprising that the emphasis in the first laboratory computer systems installed in hospitals was heavily upon data acquisition. It is now quite clear that such emphasis was not really justified. The then remarkable fact that the computer could absorb readings from several autoanalyser channels simultaneously and could register peaks almost as successfully as a junior technician was so striking that it prevented proper consideration being given to a thorough study of the laboratory functions and to the decision on which of them should be computerised first.

It is convenient to start with an examination of the laboratory's input and output. Looking at the laboratory from outside it is characterised by the repertoire of analysis it performs, the information and the specimens which must be supplied with the request and the kind of information it returns back to the requester. It is particularly the input which is affected by the environment in which the laboratory operates, and it may be helpful at this point to place the laboratory computing in the context of hospital-medical computing.

Medical computing in the hospital may be divided into three categories:

a. administrative management of patients (including outpatients), usually referred to as the registration, admission, transfer and discharge (RATD) system,

b. laboratory computing, and

c. clinical management of patients which includes all the applications involved in diagnostic and treatment processes and also research activity.

From the computing viewpoint the problems in these three categories require rather different approaches and should, therefore, be kept clearly separated. It is the role of the system interface to maintain the required independence and at the same time to allow communication between the systems, essential in an integrated hospital system. It should be noted that independence of the system does not necessarily require independence of the hardware.

The quality and quantity of information arriving at the laboratory with the specimen will to a large extent determine the system functions visible to the user on the output, existence of a RATD system, registration of outpatients and education of the users being the most important factors. The system design must therefore specify the minimal amount of information which the laboratory needs for the requested analysis to be performed correctly and the results returned to the requester. Further, it should specify what other functions can be performed if extra information is available. The system cannot for example quote sex/age specific reference ranges if the sex and age (or date of birth) are not supplied. Similarly, to attempt any kind of clinical interpretation of the results in the absence of diagnostic and treatment information is unwise and may even be misleading.

Crucial to all the applications which use cumulatively filed results is the existence and availability of a unique patient identification (e.g. a hospital number). If that is not provided by the RATD system, the laboratory may maintain a patient index file using the identification details supplied. The cost of maintaining a full 'master' index file can hardly be justified by the laboratory applications only, but it has been demonstrated by Alexander et al(2) that an index file of limited size for patients selected by some diagnostic criterion, i.e. those cases where the cumulative record is genuinely needed, may be supported relatively cheaply. Intervention by the laboratory staff will, however, be needed to resolve the ambiguities that arise when incorrect or incomplete identification is supplied. The best that can be done by the computer in such a case is to print a list of probable matches which are scrutinised and reconciled manually.

Another point which needs examination is availability of the request. The request may be delivered to the laboratory either with or without the specimen. If the request is received in advance of the specimen, the system can prepare collection lists–these may be in the form of sticky labels to identify the specimens–giving instructions about the kind of container and coagulant to be used and the total amount of blood required, among other items. As Raymond(3) pointed out, the collection list is particularly useful in a combined pathology system, allowing all the specimens to be collected on one visit of a patient. The number of subdivisions of the specimen in the laboratory can also be reduced if the collection list gives instructions about the number of containers to be used. It should be emphasised that the specimen must be fully identified since its details will have to be entered to the system on arrival at the laboratory.

Different kinds of outputs (reports, ward, doctor, clinic lists, etc.), their formats and the way in which they are distributed will characterise the output of the system. These facilities, as well as all the other functions discussed in this section, may be included in the laboratory system as applications but do not, in any significant way, affect its basic design. It must, however, be borne in mind that some of them will be a not negligible burden on the system resources and may even, if the hardware or the operating system are not appropriately chosen, overtax the system.

BASIC SYSTEM FUNCTIONS AND APPLICATIONS

The aspects of a laboratory's work upon which the computer can have most influence are, in a word, managerial; they include work simplification, record keeping, data vetting, workload monitoring and above all, the maintenance of specimen identity from the requesting to the reporting stage. It is the organisation of the work flow through the laboratory and in particular, the collation of the results of the tests requested on a patient's sample (or samples), obtained at the different locations in the laboratory and not infrequently on different days, with the patient, request and sample identification details, which is crucial to the system design. The system must, of course, perform a number of supporting functions to enable the collation process to take place, to ensure its integrity and to facilitate the input and output. It is convenient to discuss these functions under separate headings.

ACCESSION NUMBER

To assist the merger described above, details of each test request must be entered into the system and uniquely identified. It is a standard practice to allocate an accession number to each request as it arrives and to use this number for identifying the specimen, aliquots taken from it and the results of analysis made on them during subsequent stages.

The choice of accession number should not be arbitrary. Obviously, any number not used again for at least the period needed to complete the test taking the longest time, allowing for possible breakdown of instruments and the necessity to repeat the analysis, will fulfil the basic requirement. Numbers consisting of many digits with no intrinsic meaning are not popular and will be a possible source of errors (another aspect of the human interface). In general the date or some part of it (e.g. the day of the month), used as a part of the accession number is found useful, particularly if the accession number is used for filing of the requests, residual specimens and reports. An appropriately chosen structure of the accession number may also be usefully exploited by the programmer when he is designing an organisation of the file which is going to be accessed by using the number as the key. In the laboratories where the majority of the requests contain a patient number as part of the patient identification, the accession number may be constructed from the patient number, the date and the time of collection of the specimen. (Instead of collection time, a sequence number may be generated in the laboratory to allow for several requests concerning one patient, received on the same day.) Such an accession number identifies uniquely not only the request but also the patient and may be used to construct a cumulative file. If the patient number is not available, the system can allocate a 'dummy' number and the linking to the previous results must be done by some other usually more complex and less reliable techniques.

It should not go unnoticed that the system may use a truncated (or transformed) accession number for certain laboratory tasks (e.g. entry of the results) to make its use more convenient for the laboratory staff.

The way in which accession numbers are allocated also has some implications for the system operation. Essentially, the accession number may be allocated either manually or by the computer system. Manual allocation is prone to more errors since the accession numbers must be entered to the computer and if sequential numbers are used some gaps may occur from time to time due to mistakes, spoiled preprinted labels, etc. making the numbers less suitable for checking and filing purposes. Accession numbers generated by the computer are therefore preferable under normal circumstances. The major drawback of this approach is, however, revealed during, or rather after, computer breakdown. The accession numbers generated manually as part of the backup procedure dictate the sequence in which the requests must be fed in and accepted by the computer during the recovery procedure.

SPECIMEN IDENTIFICATION, WORKSHEETS

The accession number, as pointed out in the previous section, must be associated with a specific request, and containers (cups) containing aliquots of a single specimen must be linked to that specimen's accession number. The second of these problems, arising only in systems where laboratory procedures call for subsampling, is usually handled by the production of worksheets.

The accession numbers of the cups for a particular work station are listed in the same sequence as that in which the cups themselves are placed on the rack or sample holder. The cup sequence is thus vital to the correct identification of the samples since it relates results to requests.

Important facilities may be provided by a computer system which maintains worksheet files. As the test requests are input, consecutive cup numbers are allocated to the successive samples inserting in the appropriate places drift standards, quality control specimens, etc. These files can be printed out as required; the printed worksheets provide loading schedules for the sample holders and can also be used for recording the results of tests, especially manual ones, if direct data entry is not available.

Most importantly, however, it is also possible to print small adhesive labels each bearing an accession number, a position (cup) number for a particular analyser and a test code as described by Vitek et al(4). If these are printed without delay as the worksheet files are updated they can be used by the technician preparing the aliquots and also to set up sample holders without waiting for complete worksheets; they also provide positive identification of the cups throughout the analytical procedures. It should also be noted that the worksheets do not have to be printed out at all if the results are converted into a machine-readable form automatically. The use of the worksheet file for entry of results is discussed below.

Intricate arrangements are usually needed to insert emergency specimens in the middle of the worksheets set up for routine work. The use of special equipment, limited number of requested analyses and the speed with which the laboratory must act are some other reasons in favour of processing the emergency requests by a special emergency laboratory. The results, suitably marked, may then be entered to the computer system for filing or the analyses repeated by processing the request again in the routine way.

CODING OF TESTS

For the computer system to be able to maintain the worksheet files and to print out the correct number of sticky labels all the tests performed in the laboratory must be defined to it. Inclusion of all possible tests is highly desirable since it simplifies the work in the reception and eliminates another likely source of errors. Flexibility of the system will be significantly enhanced if the tests are defined on several levels.

Normally, there will be a code associated with each work station (e.g. multi-channel analyser, manual test, etc.) for which a worksheet might be generated (work station level).

If more than one determination is performed at the same work station, each determination must be separately defined if the system is supposed to handle its results independently (determination levels). It may, for example, be required to enter or delete results of one test normally performed on a multiple analyser if one channel fails. Another, more controversial reason is to 'allow' a user to request just some tests from a battery of tests and return to him only the results he requested. That facility, without arguing for or against it, may be provided by the system quite simply without affecting the laboratory procedure.

Sometimes, several tests which are not performed on the same piece of equipment form a diagnostic group or profile and are frequently requested together. It is convenient not only for the data preparation staff but also for the requester to define such tests as a group (it may include further groups as well as single or multiple tests) with its own code (group level). Independence of the individual constituents of the group must not, in this case, be affected in any way, i.e. it must be possible to request them separately or even include them in several groups.

ENTRY

Results of the analyses after they have been collected processed and validated must be merged with the request records. The way this is done will depend on whether a worksheet file for the particular list exists or not.

If the worksheet file exists, the results may be input in cup number sequence after the test code and the cup number of the first result have been entered; obviously, a new cup number must be entered every time the sequence is interrupted. An accession number stored in the corresponding cup number position on the worksheet provides an access key to the appropriate record.

For tests without worksheets, each result must be input together with the appropriate accession number and the test code. This approach may also be used for tests with worksheets if it is more convenient for the laboratory staff to supply results of requested tests grouped by requests rather than by tests. In this case an accession number will be followed by a variable number of test code result pairs.

DATA ACQUISITION

There are a number of different ways in which analytical results can be input to a computer system, and the relative merits of these will depend to a large extent upon the analytical equipment producing the results. The different data acquisition techniques will not be discussed here since they do not directly affect the system design and in fact, it is an important principle to be observed in the designing process to keep the data collection and the data processing modules clearly separated from the basic system structure. The format of data and the communication protocol required by the system will be included in the interface specifications. This design philosophy is in good agreement with and greatly assisted by the technological developments in the field of microcomputers with the associated tendency towards dispersal of the processing tasks.

DATA VALIDATION, QUALITY CONTROL

An essential function of the computer system is to validate all the information it accepts as thoroughly as possible. The validation procedures must be well synchronised with the laboratory tasks so that no unacceptable delays occur and devised in such a way that the laboratory can fulfil the request if at all possible, e.g. the specimen should be analysed and the results returned even if the name of the patient was misspelled or the patient's number was not supplied; the system must not, however, file the results automatically on a cumulative file. The data should, therefore, be validated in several stages, criteria for format and logical consistency validation becoming progressively more stringent.

Validation procedures designed to ensure correctness of the results need special attention. Individual results can be compared against plausibility limits to guard against gross errors which may be introduced during transcription or data entry stages. More sophisticated biological consistency checks—comparison against a recent result obtained on the same patient—may be made if an access to the previous results is fairly rapid.

Validation of the whole batch of results has two aspects. A scheme designed to ensure that a particular batch is acceptable in that it is based on correct calibration standards, is not subject to excessive drift, etc. and that the results may be released from the laboratory, is best described as short-term quality control. A carefully designed data acquisition routine, whether on or off-line, can readily incorporate suitable tests of this kind. The second, long-term aspect of quality control is concerned with assurance that results obtained on any one day are truly compatible with those obtained a day, a week, a month or longer in the past. Detection of errors, small in magnitude but cumulative in effect is the main purpose of such a scheme, which needs rather different methodology.

REPORTING DATA PRESENTATION

It is important for the results to become 'reportable' immediately after they pass through all the validation procedures. The way in which the results will be presented should be determined by the application module according to the purpose for which they are displayed or the context in which they are presented. Essentially, a distinction may be made between data display for internal (laboratory) and external (medical) use; the laboratory reporting must be designed in conjunction with the laboratory procedures in which it is going to be used but the way the data are presented to the users, should largely be specified by them and it must be expected that changes may serve as a good example here. The technician will use the previous results to check and comment on the current result and possibly to decide on further relevant analyses to be performed, whereas the clinician, after considering other related information at his disposal, will use it chiefly for making diagnosis and monitoring the treatment. It is very likely that different formats of cumulative data presentation will be called for in those circumstances. The value of cumulative reporting and the reasons why it should be regarded as an application function of the laboratory system were discussed in more detail by Vitek and Healy(5).

DISCUSSION

From the system design viewpoint all the functions which the laboratory computer system might undertake can be divided into two categories—first, the functions directly concerned with organising the laboratory's work flow which will determine the basic system structure, and second, the application functions which, under the control of the basic system, are concerned with tasks like data collection, reduction and presentation, production of management statistics, stock control of reagents, etc. The merit of designing the application functions to be independent of the basic functions is that it gives a large degree of flexibility and a clear system specification which allows gradual implementation.

Experience has made it abundantly clear that installing a computer in a laboratory can easily create more problems than it solves. There are many examples of systems which have had to be removed because the laboratory's performance deteriorated to an unacceptable extent. No sure recipe for success can be given, but it is possible to advocate an approach that may make failures less likely. This is to consider first what needs doing, and only then the methods of doing it.

REFERENCES

1. Jackson, MA (1975) *Principles of Program Design,* Academic Press Inc. (London) Ltd.
2. Alexander, MK, Corbett, MT and Reed, RW (1977) *J. Clin. Path., 30,* 356
3. Raymond, S (1973) *Comput. Biol. Med., 3,* 325
4. Vitek, P, Chiu, S, Healy, MJR and Lucas, D (1974) *Ann. Clin. Biochem., 11,* 86
5. Vitek, P and Healy, MJR (1974) *A Note on Cumulative Reporting,* available upon request from MRC, CRC, DOCAS Watford Rd, Harrow, Middx HA13UJ UK

APPRAISAL OF MANUFACTURERS' BASIC SOFTWARE

L. Forth, MSc, BSc
Lecturer, Department of Computer Science, University of Essex

INTRODUCTION

The role of the minicomputer has changed considerably since its inception over a decade ago. Created as a tool for process control and instrumentation, it initially fulfilled the need for an adaptable controller which could be situated at the heart of a complex technological system. During these early years, progress in other fields of application was slow largely due to inadequate system software; systems which were operational, generally had specialised software written for them in assembler code. The steady reduction in prices of the hardware together with improvements in technology in both processors and peripherals encouraged the development of more sophisticated software. This took the form of operating systems, editors, macroassemblers and problem-orientated high-level language compilers, which have ensured the success that the minicomputer has subsequently enjoyed.

The development of the single chip microprocessor unit is significant because it brings the hardware costs of computing down yet again, thus enabling low budget applications to come within the scope of the microprocessor unit. The challenge made by the microprocessor unit to the present type of minicomputer must eventually succeed, but potential buyers should not expect too much from these newcomers, firstly because their effective computing power is not as high as the established minicomputers and secondly, because extensive software support is presently not available. Thus, microprocessors currently extend and complement the areas of application of minicomputers, the former being aimed at the low-cost end of the market. Direct competition only occurs where a microprocessor has an instruction set and an input/output architecture both of which are identical to those of an established minicomputer. These units are then able to utilise directly software originally written for the minicomputer; indeed a number of manufacturers have taken advantage of this fact by producing in-house microprocessor units themselves.

The selection process for any computer system will naturally take into account the differences between hardware. Unfortunately, the merits of the software are not so easily assessed, the unwary falling easy victims to commission-seeking salesmen. Experienced buyers spend a great deal of time endeavouring to get a realistic picture of the performance of manufacturers' software. First time buyers often do not fully realise how important is the matter of software, nor how expensive will be the remedies for deficiencies which subsequently come to light, and for which they have no comeback on the originator.

SYSTEM SOFTWARE

The establishment of a new computer system should be an organised and well planned operation(1), one part of which is to produce a software specification for the system. This specification will almost certainly be modified following contact with potential suppliers, whether this is a computer manufacturer, an independent software organisation, or members of one's own organisation. It is, nevertheless, an essential document since it represents the design objectives of the system.

Software is expensive; undoubtedly the best value is the manufacturer's software supplied with the machine, provided it meets the requirements of the software specification. If this software is not ultimately to be used, it may still be necessary as a development system during the writing and testing of the target software. Manufacturers' policies do vary with regard to charges for standard software, ranging from £0-2,500 for an operating system and £0-1,000 for a high-level language (e.g. BASIC). It does represent an area for bargaining between customer and salesman depending on the size of the overall contract. But against these charges the current rate for independently produced software will be high, since the national average for finished and documented code is 10 statements per man-day.

The trend towards cheaper hardware has led to a significant expansion in operating systems (OS) over the years. For many of the minicomputer manufacturers this has resulted in several different OS versions being available for the same machine. The OS, for most purposes, can be considered as an integral part of the machine since its job is to handle all the resources of the hardware, systematically schedule the tasks created, control their execution and make communication easier with the applications programmer.

The early OS versions did little more than provide a disc based filing system for program development. These have been improved, to the point where control of all run-time resources has been included. Operations are now often carried out with reference to a real-time clock, making it possible for the events in an application to be timed and handled by interrupt programming within the OS. These disc operating systems (DOS) or real-time operating systems (RTOS) are usually the smallest of the family in size, typically requiring from 8K to 16K words of core store in the minicomputer. They generally only permit one user program to be in operation at any one time.

The desire to utilise slack central processor time, when slow peripherals are being operated or when the central processing unit (CPU) is in an idle loop waiting for a real-time event, has led to the development of the OS which permits the simultaneous running of both a foreground and a background job. This allows a machine with a real-time program operating in the foreground area also to run a batch job where timing is not important; for example, program development might be carried out in the background while the foreground program controls the task to which the machine is principally dedicated. Quite naturally, this development imposes bigger demands on the computer store, although any increase in size is sometimes minimised by complex swapping between core store and disc store. More usually, the system designers prefer a larger core store, since more straightforward operation with better protection between jobs can be obtained by strictly partitioning the available space into foreground and background areas. Inherent in this type of OS is the notion of multi-tasking, the method by which the OS controls the concurrent operation of the elementary parts of either one or more programs. Core store requirements for systems permitting partitioning are generally between 24k and 32k words, the larger size often being desirable in order to give adequate working space to both foreground and background jobs.

More recently, advances in OS design have taken advantage of hardware changes. Since most 16 bit minicomputers have an addressing limit of 32k words of store any further increase in OS size would have severely curtailed the user job space. But now, extra storage modules can be added in 32k word increments, whose management is carried out by hardware registers, containing pointers, which effectively allocate these areas to a given program. Such a system of virtual addressing enables a minicomputer to rival main frame computers by now offering larger core space and better protection not only between jobs running in foreground and background partitions, but also between multiple tasks running in the same ground. Mapped memory units have also led to multiple programming operating systems being offered with a number of machines, this being an extension of the number of grounds or partitions within any CPU. Each program is run in a separate partition and is completely independent of any other.

If any note of caution needs to be sounded in connection with the development of the minicomputer OS it is that for many applications it is overdeveloped. Computer manufacturers have over the years been moving their product up-market, aiming at the business end of the spectrum with a variety of commercially-orientated products. In the absence of clear objectives from the buyer, salesmen tend to impress the larger systems on customers with the presumption that larger is better. It is frequently the case that these advanced operating systems carry overheads which at best are unnecessary, and because of their inefficiency, may indeed be unacceptable. In many cases, the simpler operating systems would suffice, reducing in turn the expenditure on processing power and memory. In those applications requiring multiprogramming, the buyer should investigate carefully whether or not separate processors would be more economical and beneficial in the long run. The major expenditure, after all, is in peripherals and software; these facilities can often be shared between processors(2) giving benefits in CPU reliability (by duplication) and savings in peripheral costs, either by purchasing from independent manufacturers or utilising already existing configurations(3).

Concern for the efficiency of operating systems frequently leads customers to take a do-it-yourself approach, either by modifying an existing operating system or by writing a purpose-built package. In the case of modification of the operating system the essential source language listings may be withheld by the originator. Even when listings are available, the source language is most frequently assembler code, making the understanding of the way the operating system works difficult to perceive, and alterations a protracted and error-prone affair. It is highly desirable that an operating system is written in a high-level language; unfortunately there are very few examples, particularly from computer manufacturers. Many of those in existence have been written by independent organisations (4, 5). Some of the benefits of an operating system defined in a high-level language are that it is modular, self-documenting, readable and lends itself to easy modification.

Writing an operating system from scratch is viable and economic provided that a high-level language is used. Indeed the availability of a suitable language may help to decide which machine is selected for a particular purpose. Some of the requirements of an operating system are difficult to provide in a high-level language unless it has been designed specifically for system writing. BCPL, for example, exhibits a mixture of high-level and low-level language features. Alternatively, the use of two languages which can be separately assembled and loaded together can provide the mixture of requirements (e.g. extended ALGOL and assembler code).

Some manufacturers provide a skeleton real-time operating system which is usually core-based and aimed at the process control environment. It does not support disc resources but contains an efficient multi-tasking scheduler for the support of a number of user-applications tasks together with a despatch table whose priorities can easily be set or altered. Whilst not giving the all-embracing features of the larger systems, this compact system is useful in certain applications; for example in laboratory situations, where ease of modification might be important.

The operating system of any machine is probably the most difficult program to assess in the whole suite of supplied software. New systems should be avoided, since they invariably require a settling down period during which mistakes come to light and are eliminated. In addition, the frequency of issue of major revisions is a matter of concern throughout the whole of the industry, particularly as customers have to pay for each major revision. Contact should always be made with a number of users to find out what the snags are *before* purchasing an operating system; no self-respecting salesman will refuse to supply the names of previous customers. Alternatively, a visit to a user-group meeting can be useful in making contact with other users, and provide an opportunity to hear the problems of particular systems.

LANGUAGES FOR APPLICATIONS

Most minicomputer manufacturers provide facilities for writing application packages in a variety of languages. Unfortunately in the past, there has been too much concentration of effort in assembler code programs both inside the computer industry and by independent writers of application packages. This has retarded the progress towards efficient high-level language compilers. The usual reasons given for assembler code programming are that the resulting machine code is more efficient and compact than would result from the use of high-level languages. Additionally, high-level languages are often criticised for a lack of particular features necessary for the job in hand. As previously mentioned, some systems are able to link together and load code segments from different compilers. As a means of sidestepping the disadvantages of particular high-level languages, this approach has much to commend it. It is realistic to assume that certain languages are aimed at particular jobs and these are the jobs for which they are very good. Attempts at all-embracing languages have been notably unsuccessful and inefficient.

High level languages were originally conceived as machine independent languages but it is seldom that identical programs will run on different machines for the following reasons: implementation of the basic language may differ in some minor details; extensions to the language are commonplace and, if used, are machine dependent; utilisation of operating systems calls are specific to the machine, e.g. disc file accesses or other input-output operations. All of these areas might need alteration in moving a program from one machine to another. Despite these problem areas, the benefits from the use of a high-level language are also high, these include: reduced development time, improved readability and comprehension, good structuring and modularity, easy modification of the code, self-documenting.

The properties of the various high-level languages are well documented and therefore only a summary, for the benefit of newcomers to the subject, is shown here.

1. FORTRAN 2. ALGOL	Scientifically orientated batch programming languages	
3. COBOL	Commercially orientated batch programming language	
4. BASIC	Scientifically orientated interactive programming language	
5. CORAL(6) 6. RTL/2(7)	Real-time programming languages	
7. PASCAL(8)	General purpose scientific and system programming language	
8. MODULA(9) 9. BCPL(10)	System programming languages	e.g. process control e.g. compiler writing

It is worth noting that many of the implementations of the first four common languages frequently include powerful extensions to the basic definitions. These are: string handling, array or matrix operations as well as complex file handling routines. BASIC is the best established interactive multiterminal language, but is frequently only available in an interpretive form which demands the use of the whole machine. One or two examples are becoming available where both interpretive and compiling systems can be bought, these now running under an advanced operating system, which permits other non-BASIC jobs to be run at the same time. The popularity of these systems has also led some manufacturers to offer derivatives of COBOL for on-line multiterminal work, mainly aimed at commercial users.

Items 5 to 9 in the list are more recent developments which illustrate the encouraging trend to overcome the last bastions of assembler language. Since they may be less well known, references are given.

MAKING AN ASSESSMENT

One method of assessing manufacturers' software (and hardware) often mentioned, is that of benchmark testing. How feasible or practical this might be can only be judged by those people concerned in the evaluation. If individual relevant tests cannot be made, then general information can be abstracted from tests into the efficiency of minicomputers(11,12) and microcomputers(13,14), carried out by independent and commercial organisations. Great care should be exercised in extrapolating results since benchmarks frequently are only a general guide to performance and are not usually representative of the real load.

Of paramount importance for any system is the choice of disc store. Modern systems can spend up to 50 per cent of their time engaged in transfers between disc and core store; part of the time is spent seeking access to library subroutines and part in file transactions. Usually, the larger discs have higher data rates which will substantially affect the speed of operation of an entire system. Unsuitability in the available file structures for particular applications can lead to further inefficiency. Efficient utilisation of disc space can be achieved by random file organisations, but this needs to be balanced by alternative structures aimed at quick access such as sequential and index-sequential organisations.

The provision of particular high-level language compilers is often of major importance in the selection of a computer system. It is, however, the unexpected problems which tend to confound their users long after the system has been commissioned. It should be established that a large degree of compatibility exists between the OS and a number of languages to enable complex programs to be developed. As mentioned earlier it is often necessary to resort to writing packages in different languages, the whole to be linked and finally loaded together.

For larger programs, it is essential that compilation of separate source language segments can take place without reference to the remaining segments. To make this possible, compilers should have the means of indicating that certain names declared in the current source unit may be referenced by another unit; also, names used in the current unit are declared elsewhere. These facilities are sometimes known as 'externals'; a separate type distinction is often established between externals which reference data types, and those which reference relative addresses for use in an instruction. By such a distinction, better checking and fault reporting is possible during the linking process when these externals are resolved.

Another desirable feature for inclusion in a compiler is the means to insert source sections by a substitution command. A typical example is the GET 'FILENAME' directive in BCPL. This enables source code which is to be used in a number of different places to be inserted from a single common library source, specifically associated with the compiler. Each compiler should also be capable of receiving input streams from a number of files without the necessity of indicating within each file whether any other file is required. This feature is often best provided as an OS option at the command line interpreter or job control language level.

The output of compilers is generally in the form of relocatable binary files; a valuable aid, however, are optional listings in assembler code. This makes it an easy matter to check the efficiency of a compiler for any operation and also provides the opportunity of assessment before purchase.

Library routines, already in relocatable binary, are often provided by a manufacturer. It is necessary for efficient program development that easy modification to the system library structure is possible for users wishing to change the library, perhaps by adding routines of their own. This ensures that only single copies of well used routines are in existence, ultimately saving the space caused by unnecessary duplication.

Some features will be dependent on the type of hardware employed; for example, those machines which have multiple base and limit registers. In such cases, it will be possible to allow programs to make reference to more than one data segment, selection being by relocation by hardware. It is clearly necessary in these sytems to be able to define which segment is being selected for use, especially where segments are shared between different processes.

Other traps awaiting the unwary include limitations in working table lengths used by the compiler, which affects the number of separate procedures and externals which may be used. Stack depth limitations control the number of nested routines which can be handled by a compiler. Finally, the addressing structure of the machine may be an embarrassment to the production of larger programs because small addressing fields limit relative movement or data reference, without the large scale use of indirect addressing techniques.

CONCLUSION

There is no doubt that small computers have enormous potential and will find greater acceptance in many more applications than has been the case up until now. Where a computer is considered suitable for a particular application the best possible evaluation of hardware and software should be carried out.

Nothing should be taken for granted in determining the best supplier of equipment. Do not accept the word of the sales force; to do so is to court disaster. Do talk to other users of similar equipment and learn by their experience. Satisfy yourself that there is adequate support in this country if the supplier is overseas.

Good planning, observation of others and a commonsense approach will ensure success.

REFERENCES

1. Wilson, DL and Webb, MJ (1976) In *Mini-computers and Small Business Systems, Online Minicomputer Forum.* Page 513.
2. Forth, L, Kidd, PA, Lyons, DM, Bowden, KF, Corker, SW and Strutt, C (1975) *Proc. I.E.E., 122,* 785.
3. Forth, L, Kidd, PA, Corker, SW and Strutt, C (1976) In *Minicomputers and Small Business Systems, Online Minicomputer Forum.* Page 155.
4. Ritchie, DM and Thompson, K (1974) *Comm. A.C.M., 17,* 365.
5. Brinch Hansen, P (1976) *Software-Practice and Experience, 6,* 140.
6. Woodward, PM, Weatherall, PR and Gorman, B (1970) *Official definition of CORAL 66.* HMSO, London.
7. Imperial Chemical Industries Ltd.(1974) *RTL/2: Language Specification*
8. Wirth, N (1971) *Acta Informatica, 1,* 35.
9. Wirth, N (1977) *Software-Practice and Experience, 7,* 3.
10. Richards, M (1969) *Proc. AFIPS, SJCC, 34,* 557.
11. *Minicomputers and Europe,* (1972) Report of the Electrical Research Association, Leatherhead, Surrey.
12. Wootton, J (1976) *Minicomputers and Small Business Systems, Online Minicomputer Forum.* Page 343
13. Penney, BK (1977) *Conference on Computer Systems and Technology, IERE, 36,* 63.
14. Lee, MD (1977) *Conference on Computer Systems and Technology, IERE, 36,* 73.

ASSESSMENT OF FILING SYSTEMS

I D P Wootton, FRCP, FRCPath, FRIC
Department of Chemical Pathology,
Royal Postgraduate Medical School, London W12 0HS

INTRODUCTION

In most clinical laboratories, the scientific equipment which is used to examine the specimens is more highly developed than the data handling system which deals with the results. In consequence, a disproportionate amount of the laboratory time is spent in clerical duties and much of the laboratory effort is potentially wasted because results cannot be reliably and quickly recalled. In addition, errors in transmission or transcription are comparatively frequent and these 'random mistakes' have caused concern recently(1).

The natural solution to these difficulties is to use a computer based data processing system. When computers were initially introduced into pathology laboratories, many of the biochemistry analysers were of the continuous flow type with an analogue output which required extensive processing to convert the readings into usable results. It was therefore natural that considerable attention was paid to the matter of 'peak picking' and calculating the results from the peaks(2). In particular, installations were established in which there were a large number of continuous flow analysers working in parallel, with the computer multiplexing the analogue results. Raymond(3) has stated that until recently the great majority of the computer systems in America were 'on-line real-time direct data acquisition systems programmed on the naive assumption that arithmetic is the most critical problem facing the laboratory. It is not; it is only a small part of the problem'. In fact, we can anticipate that the data acquisition problem will become less pressing in the future than it has been in the past, with the evolution of new equipment. It is commonplace now for modern apparatus to contain sophisticated computing equipment in its output stage, and the development of microprocessors is particularly important in this connection. Any manufacturer must take account of the majority of laboratories who do not possess data processing systems and he is therefore bound to provide his new apparatus with the ability to calculate its own results.

We therefore conclude that the main computing problem in the clinical laboratory today is a data management function which can be split up into several well defined tasks common to all clinical laboratory disciplines. These are, firstly, test requesting which is the registration of requests for tests received by the laboratory and includes details of the identity of the patient as well as the specification of the test or tests required. Second is file handling and maintenance which is concerned with collecting together, sorting, validating and storing the results of the laboratory. Third is reporting the results, which involves the collection of the necessary information from the files and its output, usually in the form of a printed document but often also as a visual display. All of these functions are critically dependent on a well designed and efficient filing system.

THE FILING PROBLEM

The working data store in a laboratory will comprise the test requests which have been received in the more or less recent past, both those which have been completed and those which are still being processed. This is the main body of data to which the laboratory requires access and this discussion will concentrate on it. In addition to this clinical file as we call it, there will be other subsidiary files which are mainly used for internal laboratory purposes, such as work files, report files, etc., which are necessary for the operation of the laboratory in producing reports etc.

It is assumed that this clinical file is in a dynamic state, being continuously replenished by the daily input of fresh requests. Since the physical size of any file is necessarily finite, a routine of culling the file by periodically removing the oldest records must be adopted. The culled records can be stored in archival files; they are required less frequently and hence the access speed is less important.

The Information to Store

In the clinical file, the unit of information can be regarded as the result of a single test. Pathology samples are received in the laboratory with requests that one or more tests be performed on the sample; the request can thus be regarded as a set of tests where the number of tests in the set may be one or more. Further, on behalf of each patient there may be more than one request, thus the patient may be regarded as a set of requests where the number of requests in the set is one or more.

In conventional manual laboratory files, there appear to be two main ways in which the results can be stored. The first, which is common when copies of the reports are kept in the laboratory as the laboratory record, uses the completed request as the filing record. This has the advantage of providing a relatively small record unit with the corresponding advantages as far as allocated storage space, etc. is concerned and means that there is no necessity in the laboratory to link today's requests with those originating from the same patient on previous days. Thus a laboratory commitment to establish a unique identity for each patient is avoided.

The other main file structure used has the file arranged as a hierarchy, so that the information is sorted by patient, within a single patient's record by request. Such a structure makes it relatively simple to recall the complete records of an individual patient; it thus lends itself to cumulative reporting and to interrogation, e.g. by visual display unit, when all the data about a given patient can be reviewed. The purpose of the file is to provide assistance in the clinical management of the patient and the clinical value of the information is much enhanced by reviewing each test result in the context of others performed on the same patient. It would therefore seem that storage of results in patient order would have significant advantages.

There is good agreement between different laboratories on the information which must be stored. This can be considered under two headings, firstly the patient identity data and secondly the test request data. As an example, the clinical information stored in our own system, the Phoenix system, is as follows:

Patient identity record
- casenumber
- surname
- forename
- date of birth
- sex
- hospital
- ward
- consultant

Results record
- laboratory accession number
- specimen type
- date of collection
- test and result (several if necessary)
- clinical information supplied (not always present)
- free-form comment (not always present)

Other laboratories store broadly similar bodies of data and the closely similar request forms designed by most hospitals indicate that different laboratories have the same requirements.

Filing Operation

It seems self-evident that the clinical file must be capable of allowing random access. The clinical laboratory has no control over the requests received by it on any particular day and they are therefore, from the laboratory point of view, a random sample of all the patients in the laboratory's practice. Further, the tight schedule of receiving specimens, testing them and reporting the results promptly means that any batching and presorting of requests is liable to impose an unacceptable delay. Successful systems are therefore those which have efficient random access files. It is true that systems have been devised in earlier times based on magnetic tape handlers with the necessary consequence that they were confined to sequential files only. The problem was tackled in these early systems by batching the information, sorting it and then accessing the sequential files in the correct order. A notable example of this type of system was the Elliott Medical 903 A system, which was certainly successful. However, the workflow which it could manage was strictly limited and much time was spent on sequential reading and rewinding of the magnetic tape files.

The amount of traffic on a filing system in the laboratory is often underestimated. Our own system deals with a workload of about 500 requests per day; the clinical file, however, is accessed 65,000 times each day (this is an actual figure obtained from a count included in the system programs). Many of these accesses can be accounted for by various internal housekeeping operations on the files but even when the latter are excluded, at least 15,000 records are written or read each day for laboratory purposes. It should of course be understood that a single request will require multiple references to the file, correctly to locate it in its position, to update it as results are available and to interrogate it for the production of worksheets and other documents and to produce the laboratory reports.

The very large number of file operations emphasises the great importance of a fast access time. In our own system, the access time is less than 1 second and thus under 4 hours of 'system time' are required for direct file accesses. However, should a filing system be used which has an access time of more than 1 second to read a random record, it would become unmanageable. Table I indicates how rapidly this would come about.

Table I

Average time of access	*System hours used for file operations*	*System hours available for other purposes (a)*
1	4.2	35.8
2	8.3	31.7
3	12.5	27.5
4	16.7	23.3
5	20.8	19.2

(*a*) calculated on the basis of an 8 hour day and an average occupancy of five programs in a time-sharing system.

A POSSIBLE SOLUTION

Since we require both to read and insert records in a random order, there appears to be two file structures which are suitable. The first is a chained structure whereby the head of each chain is occupied by a patient identity record which points to the first request record of that patient. Each request record then points along the chain to the succeeding one. New request records are added to the first vacant position on the file and are linked to the last request for the same patient(5).

This relatively simple file structure has many advantages when the length of the chain is short. Unfortunately, with the increase in the amount of laboratory work being performed on each patient and with the likelihood nowadays of associating laboratories in several disciplines together in one data system, the length of the chains tends to become unacceptably long so that such a filing system has not the very rapid access which is required.

The alternative and more successful method is that known as 'indexed sequential'. This combines efficient sequential access to data with fast access to individual records by a key and allows insertion and deletion of records.

Each record must have an internal label or key by which it can be recognised by the filing system. In the case of clinical records, the obvious key is some unique part of the patient identity record and the patient's hospital case number is usually selected for this purpose. By definition, the case number is allocated so as to provide a unique identification of each individual patient and this gets over the difficulty of identical surnames and other data. However, although in an ideal world the case number will always be available when the record is required, in practice the file must frequently be interrogated knowing the patient's name but without knowing the case number. Subsidiary arrangements must thus be made to enable such a search to be initiated and this involves either a secondary key consisting of part or all of the patient's name or alternatively a subsidiary file must be maintained to provide a link between the name and the case numbers associated with it.

The indexed sequential filing system supported by Computer Technology Ltd and used in our laboratory has only a single key in each record. The records are stored in file divisions called buckets. Records are packed into one end of a bucket and their actual position in the bucket is recorded by pointers to them packed into the other end. Within a bucket, records may be stored in any order, but the pointers are always maintained in strict key sequence so that accessing records in the order of their pointers gives sequential access.

Random access is obtained by employing indexes. The indexes do not contain an entry for every record in the file, but list the highest key that may be stored in each bucket. Because records are stored in key sequence throughout the file, the indexes can be used to locate the bucket in which any particular record should be stored. Two levels of index are used. Adjacent buckets are grouped together and there is a group index. For each group there is a subsidiary bucket index. As a result, although our clinical file holds about 50,000 records, any single record can be retrieved in the time required to do two or three disc reads.

The free space remaining in a bucket allows for the insertion of new records and is always kept as a continuous area and not allowed to become fragmented by the insertion or deletion of records. At file creation, free space is left in each bucket so that the file can grow. However, as they are introduced, new records will not be distributed evenly throughout the file so that the space in any particular bucket may easily be exhausted by a large influx of new records to that bucket. When an incoming record cannot be accommodated in the bucket to which it belongs, it is placed in an overflow bucket. Even though there is not room in the original bucket for the new record, a pointer to it is always inserted in the correct place which indicates the overflow bucket into which the record is actually stored. Access to a record in overflow is clearly slower than that of a record in normal storage, so that from time to time the file must be tidied to restore its normal structure.

In the Hammersmith Phoenix system, the case number is the most significant part of the key(6). The least significant part of the key is the sequence number of the request relating to the patient; by combining the two parts of the key in this way, all the data referring to a different patient is kept as a contiguous section in the disc file. This is convenient when it comes to operations in which the patient's data must be treated as a whole. This applies for example to the preparation of cumulative reports or to visual display unit (VDU) enquiry on the results of the patient.

DISCUSSION

This description of the features of a good laboratory filing system assumes that the laboratory should be able to look up the work in progress and the work which has been completed in the recent past. If this facility is not wanted, then no filing system is really necessary as the work can be put through on a basis that each single request is dealt with separately and generates a single non-cumulative report. It is, however, quite clear that the ability to relate together the results of several requests means that very much more value can be extracted from the results and the establishment of an efficient file is a very worthwhile objective.

Although computer systems have been installed in some clinical laboratories for more than 10 years, it is only comparatively recently that adequate clinical filing systems have been introduced. In contrast to our early expectations (7) it has taken a considerable time to solve the technical problems. There would seem to be two main reasons for the delay. Firstly, much effort was expended on the development of data acquisition from laboratory instruments, a function which as explained above is likely to become obsolete with the evolution of new instruments. Secondly there has, I believe, been a general underestimation of the amount and complexity of the data management required if the computer system is to replace (and improve) the functions performed manually by staff of a busy laboratory. This has resulted in attempts to solve the problem with inadequate hardware or software (or both) and has certainly lead to a feeling of disillusionment in some quarters. Successful systems now in operation show that this reaction is unjustified.

There are two diverse concepts for the laboratory data processing of the future. On the one hand, the advent of cheaper, small computer systems could make it possible for many laboratories to install their own independent data processing, using locally held files as indicated here. On the other hand, many computer scientists believe that we should take advantage of the economies of large scale operations and that the laboratory should become simply a peripheral to a main hospital data bank. It is of course attractive to a laboratory worker to foresee the time when his responsibility for maintaining the laboratory files will be accepted by another department. However, the large number of internal transactions revealed by our own experience indicate that this solution may not be as easy to achieve as is sometimes thought and that probably a considerable proportion of the laboratory processing would be best done locally, perhaps with transfer of finished data only to the main hospital files.

ACKNOWLEDGEMENTS

It is a pleasure to acknowledge the contributions of the computer staff and my laboratory colleagues at Hammersmith in developing our system. We are much indebted to the Department of Health and Social Security who provided the computer and support our activities.

REFERENCES

1. Whitehead, TP (1977) *Quality Control in Clinical Chemistry.* John Wiley & Sons, New York, London, Sydney, Toronto
2. Flynn, FV, Piper, KA and Roberts, PK (1966) *J. clin. Path., 19,* 633.
3. Raymond, S (1974) *J. Am. med. Ass., 228,* 591
4. Abson, JA, Prall, A and Wootton, IDP (1977) *Ann. clin. Biochem.* (in the press)
5. Judd, DR (1973) *Use of Files.* Macdonald, London
6. Abson, JA, Prall, A and Wootton, IDP (1977) *Ann. clin. Biochem.* (in the press)
7. Association of Clinical Pathologists (1968) *J. clin. Path., 21,* 231

CHOICE OF LANGUAGE

P. Hammersley BSc, MA, FBCS(Hon)
Head of Computing Services, Middlesex Polytechnic, London
Late Director, Computer Unit,
Addenbrooke's Hospital, Cambridge

INTRODUCTION

The objective of this paper is to discuss the criteria which affect the choice of a computer language for the implementation of a computer system for a clinical laboratory. In seeking to achieve the objective it looks first at the principles behind the design of computer systems and the principles behind the design of computer languages in order to see how these react. Involved here are considerations such as accuracy, efficiency, versatility, security and transferability. It then considers the effects of choice of hardware on the choice of language or, alternatively, the restrictions which the choice of one particular language may place on the final choice of hardware. It is hoped that through the discussion, guidelines may emerge which could help future designers.

SYSTEMS DESIGN

Design for whom

The function of a language, whether a natural language, a mathematical language, or a computer language, is to effect communication. In the case of the computer language this communication is either between the systems designer or programmer and the computer or between the user and the computer. In the former case the objective is to instruct the computer as to what process it is to carry out. In the latter case it is usually to enable data to be entered into, processed by, or extracted from the computer, although languages which have been designed expressly for the purpose of enabling naive users to write programs of instructions do exist. One criterion which determines the choice of language is, therefore, who is to use it.

Design for what

Computer systems in clinical laboratories are very varied in their scope, ranging from single batch jobs to complex on-line data capture and enquiry systems. Examples of simple systems are:

1. The input of results from hand-coded documents, possibly using professional data preparation staff, so that analysis of the results may be undertaken at a later date.
2. The preparation of labels and worksheets from coded request forms to allow automatic checking of analyser equipment.
3. The performance of calculations on single results or groups of results or purposes such as bioassays.
4. The retrieval and analysis of groups of results from accumulated records.

More complex systems involve the storage of patient records and cumulative analysis and reporting on successive results for a particular patient, the results themselves being input either by data preparation or perhaps directly from mark-sense or similar documents. The most complex systems, however, involve a substantial amount of automatic data capture; for example, patients' basic data, test requests and results from automatic analyser equipment; and also a substantial amount of automatic processing; for example, quality control of the automatic analyser equipment, cumulative reporting on test results and prompting for requests which have not been actioned.

Hence it can be seen that the choice of language is dependent not only on the person who is to implement the system but also on the type of system to be implemented. Also these two considerations are interdependent, although it is unlikely that a naive user would wish to implement a complex on-line system, nor is it likely that a professional programmer would be content to program simple enquiries from files of past results.

Constraints affecting systems design

Every computer system is constrained not only by the need to carry out the tasks which the system is designed to perform but also by a consideration of the environment within which the system is to operate. For example, in a system for a clinical laboratory it is important that data which is collected is correct, no results are lost, those who are going to use the system are not confused by any mnemonics or other unusual terminology which they are asked to use, or by having to go through tortuous procedures in order to achieve simple results, and people who should not have access to results do not gain such access.

Again there are constraints which are imposed by the fact that no system should be more expensive to create and/or run than is really necessary. Hence it is important that the system can be modified or extended easily, that the method of programming the system is one which gives the greatest possible assurance that the programs operate correctly, and that the system can be transferred easily to other locations.

These are all constraints which affect the choice of the language in which a set of programs is to be written. Yet again they do not of themselves determine the choice of language. They too have to be considered against what the system is to achieve and who is to create and use it.

LANGUAGE DESIGN

Criteria for selection

There are in existence today at least one thousand computer languages, each created with a different objective in mind. Fortunately, for the purposes of this paper it is possible to classify these languages into three clear groups and to select some well known examples in order to highlight the criteria for selection.

Classes of language

In general, there are three classes of computer language: job control language, programming language, and data entry and enquiry language. Each of these can be used either by a naive user or by a professional programmer.

Job control languages can be either complex (cf. IBM 370 JCL), where every requirement has to be stated explicitly, or may be simple (cf. Xerox

Sigma 6), where every type of statement has a default value and two or three statements can be sufficient to run a complete job. Clearly a naive user is not going to be happy if he has to write a complicated job control sequence in order to initiate a process to allow him to input a small amount of data.

Programming languages may be low-level (i.e. in a one to one correspondence with actual machine code language) or may be high-level (i.e. related closely to a description of the problem being tackled) or may even be at some level in between; for example the intermediate code of BCPL or the macroassembly level code of computers such as the Xerox Sigma 6 or DEC System 20.

Programming languages are the ones on which most work has been done and about which most is known. The high-level languages are usually considered as being in two classes, scientific and commercial, whilst these classes themselves can be further subdivided; the commercial languages into processing languages (e.g. COBOL) and report generators (e.g. RPG), the scientific languages into numerical processing languages (e.g. ALGOL, FORTRAN), structure processing languages (e.g. LISP), array processing languages (e.g. APL), and string processing languages (e.g. SNOBOL). There are, however, a number of languages which are intended to be used for either scientific or commercial applications, PL/1 and ALGOL 68 being the best known examples. Within the high-level languages there are some which take no account of the architecture of computer hardware (e.g. ALGOL), therefore their use runs the risk of generating systems which are too slow, whilst there are others (e.g. BCPL) which are a high-level form of machine code and their use runs many of the risks associated with machine code programs. Again there are languages which are geared to a specific type of system (e.g. real-time systems, CORAL 66) or to a particular type of problem (e.g. simulation, GPSS, SIMULA). There exist even languages for defining languages. Any attempt to classify programming languages uniquely into distinct sets is bound to fail, hence a few of the more common languages will be considered below and their advantages and disadvantages in particular cases discussed.

The third class of language is the group which are intended for use in order to effect data storage, retrieval or analysis. Usually such languages are not defined specifically, nor are they given a formal name; more often they are associated with a program or group of programs (called a package) and the name which is referred to is the name of the package. Examples of these are BMD and SPSS, packages for the statistical analysis of test results, and IMS and ROBOT, packages for the storage, retrieval and analysis of data. The languages associated with the latter two packages are often referred to as query languages. Nevertheless, these languages are important since if a laboratory computer system is created solely for the storage of and enquiries about test results the interface between the user and the computer will be a language of this type.

Language types of particular importance

There are two languages, one within the third class and one within the second, which are of particular importance in the design of on-line systems.

In on-line systems, the visual display unit (VDU) screen formats, which technicians complete in order to input manual test results or to enquire about previously captured results, constitute a language of the third class.

It is important to remember this since the constraints which are applied when selecting a programming language (see above) must be applied also to the design of languages for the input of data.

A language of the second class which is equally important in this context is a language for defining languages. If it is possible to use VDU screen formats as a tool for defining VDU screen formats then it becomes possible to involve users in the design of the formats through which they are going to do their work.

Specific programming languages

This section refers to a number of languages which are available on many different makes and models of computer and for which the structure of the language is defined clearly, in many cases by an agreed international standard.

Machine code languages:

The criteria which apply to machine code languages are common to them all, irrespective of the computer to which they apply. It takes longer to write a program in machine code (such languages are not self documenting and therefore more time needs to be spent on documentation) and it is more difficult to ensure that programs are correct. Further, machine code programs cannot be transferred easily except to other computers of the same make and model. On the other hand, machine code programs can be written to do any task which the hardware is capable of performing (particularly important if one is dealing with direct data capture from analyser equipment) and they are as efficient and extensible as the structure which the programmer builds into the system.

ALGOL 60:

A language designed originally for describing procedures in numerical mathematics. Its design does not allow for any specific machine architecture, hence programs written in ALGOL 60 tend to be inefficient. On the other hand, ALGOL 60 incorporates good design principles which makes it easier to write correct and easily modified programmes than is the case with most other languages.

APL:

A language designed originally for array processing and more complex mathematical work. It uses a complex notation and would normally not be of any interest to a computer user in a clinical laboratory. Recently, however, it has been used to build simple interfaces to database management systems (see next section) and as such could be important in the future.

BASIC:

A language for writing simple interactive programs of a numerical or text handling nature. It is ideal for medical staff who wish to write their own simple statistical programs, but difficult to use for any other purpose.

BCPL:

A language which has achieved great popularity for systems programming (i.e. for writing the operating system, terminal handlers, and interrupt handlers which are essential in the implementation of an on-line system).

COBOL:

A language designed specifically for commercial data processing. It would normally not be of interest to a computer user in a clinical laboratory but there are some transaction processing and file handling packages (see next section) which expect the user to provide sections of code written in COBOL.

CORAL 66:

An ALGOL-like language designed specifically for writing real-time programs, i.e. programs to handle terminals and analyser interfaces. It has been criticised widely because, although specifically intended for work on real-time systems, it includes no features which are particularly relevant to that area.

FORTRAN:

The best known and most widely available high-level programming language. Designed originally for numerical mathematics it has been used in almost every type of application and, when combined with machine code, provides a highly efficient, readable and extensible combination.

MUMPS:

A language similar to BASIC which includes in addition a hierarchical data storage and retrieval protocol. It was conceived at Massachussetts General Hospital and intended for use by medical staff. It is much loved by MUMPS users but these normally are concerned only with simple data storage and recovery through interactive teletype terminals.

PL/1:

A language designed by IBM and intended to take over from both FORTRAN and COBOL. Although capable of producing very efficient and readable programs, it has not 'caught on' in the same way as its predecessors and it is not readily available on computers other than the IBM 370 range, particularly minicomputers.

RTL:

Another language for real-time computing. Better liked than CORAL 66 but not so readily available.

SNOBOL:

A language for string and text handling. It is not readily available, it is difficult to understand, and one requires considerable experience before one can write programs quickly. It is included in this list, however, because investigation has shown that the concepts embodied in it (text handling, number handling and enquiry processing) are exactly those needed by all classes of National Health Service (NHS) users. Should its implementation be improved at some later date then it might become very important.

Specific languages for data storage, retrieval and enquiry

Unlike programming languages, statistical packages, problem orientated packages, and packages for data storage and retrieval are not common to many computers and they do not follow agreed language standards. It is not possible, therefore, to do more than mention a number of such packages and to suggest that, when hardware is being selected, the availability of such packages should be a major factor in the selection process. Examples are:

BMD, SPSS:

Packages for statistical analysis of sets of data taken from stored results.

DBMS:

A package for the storage and retrieval of data in a structured form. Usually referred to as a CODASYL database management system it is available on most large computers under some different title (e.g. DMS1100, EDMS, IDMS, IDS). There are available also packages which allow generalised enquiries in addition to the simple storage and retrieval (e.g. IMS, ROBOT).

DECNET:

A package to allow communication between two or more computers in a distributed processing network.

RPG:

A set of packages for generating formatted reports from files of data, in some cases including simple processing as well.

TP:

A package for terminal handling.

Compiled and interpretive languages

There are two standard ways in which the computer can be persuaded to obey a program of instructions written in a high-level language. The first is to provide another program (called a compiler) which translates the high-level instructions into an equivalent machine-code program. This can then be run on the computer in the same way as any other machine code program. The second is to have a program (called an interpreter) which both decodes and obeys the program statements at the same time. Hence a statement is decoded each time it is obeyed. The interpreter must be present in the computer whenever the program is being run. Usually this proves expensive both in storage space and computer time and interpretive languages should not normally be used except for simple interactive storage and retrieval type problems. Of the programming languages listed above only APL, BASIC and MUMPS are normally run interpretively, although in the case of BASIC, compilers for the language are becoming increasingly common.

Mixing languages

One of the most common problems when using a high-level language is to find that there are sections of code which need to be very efficient (but the compiler does not translate them effectively) or that there are tasks which it is not possible to define in the language used. To overcome such problems it is common to write programs which mix sections of machine code with the statements in the high-level language. Normally it is easier to do this when using BCPL, COBOL or FORTRAN than when using any other language. With interpretive languages such as BASIC and

MUMPS it is normally not possible to mix sections of code. Where efficiency and versatility are major constraints this factor is most important.

THE INTERACTION BETWEEN CHOICE OF HARDWARE AND CHOICE OF LANGUAGE

In many cases in the clinical laboratory a new system is associated with a new computer. Hence it is important to know whether the choice of hardware might affect the choice of language, or vice versa.

With large and expensive hardware it is unlikely that there would be much restriction on the choice of language or languages to be used. It is more likely that available packages would impose the restriction. With small computers the availability of high-level languages is more restricted; there are some which offer only BASIC. It is most unlikely, however, that there would not be an alternative computer available at the same price which would offer FORTRAN and some form of real-time operating system.

As in the case of large systems, when considering minicomputers, the choice of language is less likely to determine the final selection than is the range of packages which the computer supports. On many minicomputers there is a wide variety of packages available whose use might save many man hours of programming effort. Since their availability is changing continuously, only a full evaluation at the time of selection can finally determine the choice.

One must not forget, however, that hardware eventually must be replaced. One must try to avoid choosing a language which seriously restricts one's choice of replacement hardware when the time comes.

THE WORK OF THE NATIONAL HEALTH COMPUTER SERVICES GROUP (NHCSG)

Early in the 1970s, first a NHS working party and later the Software and Programming Working Party of the NHCSG considered the suitability of various languages for use in NHS computing. In both cases the reports were inconclusive. Copies of them are, however, available.

CONCLUSIONS

The different types of user and the different types of system make it impossible to specify any single language which is of universal application in the clinical laboratory. Where a naive user wishes to write simple interactive programs using either a dedicated small computer or a terminal to a larger system, then BASIC will probably suffice. For real-time system controlling a number of terminals and automatic analyser equipment it is likely that any of BCPL, CORAL 66 or FORTRAN, linked to machine code routines, will be adequate. Where the clinical laboratory system is part of a much larger integrated hospital system using terminal handling and database packages it may be necessary to use COBOL. In both of the latter two cases it is wise to avoid the use of interpretive languages. In all the three cases a combination of FORTRAN and machine code is likely to meet any requirement, including the constraints of efficiency, accuracy, versatility, security and transferability.

In many cases, however, the implementation group is, in practice, designing a language for use by the user group. In this case the members of the implementation group must be aware of the principles of good language design and of the fact that the users themselves might wish to be involved in the specification of a tool of which they will be the prime users.

TRANSFERABILITY OF SYSTEMS

H J A Norman
Chief System Analyst,
North Staffordshire Health District

GENERAL

Transferability of systems is currently one of the major topics of discussion in National Health Service computing. It is a little surprising, therefore that so little effort has been made to gain from comparable experience in industry and commerce, for ideas of transferability were common in those kindred areas a decade ago. More important still, industry and commerce actually emerged from the stage of ideas and discussion to the stage of implementation on remote sites. This is a stage that the Health Services have only just reached, and reached via a route which is arguably not the best.

GENERAL IDEAS OF TRANSFERABILITY

Although there are obvious intermediate stages, the main ideas about transferability can be summarised as transfers involving:

i. same code, same type of computer,
ii. similar code, similar type of computer,
iii. dissimilar code, dissimilar type of computer.

These options will be discussed further. It will be noticed however that all three envisage a further machine. Arguably, at least one potential form of transferability is being ignored. That is the form of transferability whereby an application is replicated on a remote site but employing the home site computer.

same code, same type of computer

Although there are extreme examples of systems being transposed onto literally scores of identical machines (such as the American armed forces stock control scheme), it is as yet relatively rare to find in the Health Services machines bought specifically to match existing ones in order that identical and proven systems may run on them. Certainly no system has yet acquired a reputation and a reliability such that identical machines are being purchased with the sole intention of running it.

similar code, similar type of computer

In this type of transfer basically similar code would be transferred to a computer having a fair number of features in common with the parent one. Almost certainly the programming language involved will be one of the common high level ones, such as Cobol, Fortran or Algol. The computer itself will tend to be one of a range and either upwards or downwards compatible. Outside that probability is the remoter possibility of two different manufacturers offering a machine derived from a comm source. A notable and unusual example of this is the direct compatibility of IBM 360/40 and ICL System 4/50.

dissimilar code, dissimilar type of machine

It has been argued that this is not truly a transfer since what is transferred is not 'the original system'. Obviously the relationship between the transferred and the original system may vary widely, at the narrowest being almost indistinguishable and at the widest unrecognisable.

At present, however, this last is undoubtedly the commonest form of transfer in the Health Services, though not acknowledged as such. At the lowest level design ideas, conversations and programs are borrowed (usually without acknowledgement); at the widest, suites of programs and whole applications. Whether or not one dignifies the process by the title of transferability or maligns it as idea stealing, it is an everyday occurrence—which, in itself, suggests a certain ease and convenience.

EXPERIENCE FROM INDUSTRY

In my working life prior to joining the Health Service I worked for three very large companies. All had remote sites with their own computers, but all three adopted quite different approaches with different degrees of success.

Firm A

Firm A was committed to a very large production control system which it began to design shortly after I joined them. Fortnightly meetings were held at the principal site with both the central site and remote sites represented. Of a total meeting membership of around 20, about 15 represented the principal site, and as the design progressed the other sites' representatives were reduced to an even more ineffectual role.

At the end of the day the firm was left with a large well integrated system that worked excellently throughout its major plant. The other sites represented on the committee, however, were distinctly affronted by the minor role they had had allotted to them and so resistance to any transfer of the scheme was well developed. In fact, little or no formal extension ever took place, though an informal and unacknowledged transference took place.

Firm B

Firm B was multinational with its home base in North America. It had a total of eight worldwide sites, each with its own computer or computers. Buying policy for main frames was fixed centrally but complete liberty as regards smaller machines was allowed. The principal English and the principal American factories were recognised as having peculiar competence in accounting systems and production control respectively.

No rules whatsoever were laid down about transferring systems or, indeed, whether it was policy to do so or not. During a period of intense growth in data processing it was taken for granted that a search would be made for the desired expertise within the company and, given the compatibility of all company main frames, transference would usually be a matter of extra disc or tape capacity rather than trauma.

To those who doubt that a self-regulating system like this leads to any fruitful results I can only point to the inwards and outward traffic, at our site, in a two year period.

Inward i from America (real time production control)
ii from France (real time process control in laboratories)

Outward i to Australia (customer accounting)
ii to America (stock control)
iii to Canada (standard costing)
iv to another UK plant (stock control, standard costing, sales ledger).

Thus, from an extremely *laissez-faire* attitude occurred a very considerable interchange (helped by the extremely good information services existing within the group).

Firm C

Firm C, based entirely in the UK, adopted an entirely autocratic system. Various major computer centres were established with powerful main frames (roughly one for every four or five factories) and all basic systems were run there. An illustration on payroll may make this easier to follow.

In the centre where I worked, six factories (comprising approximately 45,000 employees) all did their own data preparation onto punched paper tape and this was transmitted via fast data links to an (off-line) paper punch at the centre. The payroll for each was processed and output either returned via punched paper tape up the data link or printed in the centre and returned via a security firm.

This firm did not insist on machine compatibility, and in fact a variety of manufacturers were represented within the firm. Obviously, however, there had to be standardisation on the types of data links used; and the use of punched paper tape as in input (and in some cases an output medium) was virtually standard.

The effect of this policy was that major and complex systems tended to be run in very few places, leaving the local (and smaller) machines to run a diversity of smaller applications. This concentration minimised the number of sites requiring to have knowledge of complex systems, but, on the other hand, little effort was made to incorporate very many options into the systems. Hence, the systems fitted very well on the original host site but not so well elsewhere. Frequently, major changes in clerical and procedural practice were called for on the other sites in order to accommodate systems designed without reference to their particular requirements.

From these three examples–and all three companies for quite separate reasons approved of, and encouraged, transfers of systems–some important points can be gained. In this connection, the avoidance of paths leading to failure is probably as important, if not more so, than the uninhibited pursuit of success.

STARTING POINTS

In seeking the goal of the transferable system, there are two usual starting points:

i A deliberate attempt to design, starting from a blank sheet of paper, a transferable system.
ii A desire to establish whether a system already existing is transferable.

In the first case, the chances of success are probably in an inverse relationship to the number of people or sites wishing to use the system.

The second example is altogether a more precisely formulated problem. A fully developed system is fitted (or not fitted) to a set of requirements existing in a borrower's mind. It probably does not matter that the borrower's requirements were arrived at without any particular formal or rigorous analysis.

However, both starting points envisage substantially similar steps in the analysis of the transferable system.

THE CHARACTERISTICS OF THE TRANSFERABLE SYSTEM

Does it do what I want?

If part of a design committee, the potential user will no doubt have assured himself of this by successive meetings and discussions. No doubt he will have, or have had, flowcharted his existing manual procedures and discussed with staff all man/machine interfaces. He will not have committed himself to a transfer until a specified period of completely troublefree running has elapsed somewhere else. Any necessary procedural changes will have been practised beforehand. A plan will exist for file take-on, and what to do in the period where both manual and computer files exist.

If the system is fully operational and troublefree already then a full and rigorous examination of the options and facilities within it should be carried out. Particularly is this the case where the current operation appears to be nearly, but not quite, what is required. If the system is parameter driven, or contains its own dynamic options, then a closer fit with requirements can be achieved relatively cheaply.

In particular, great care should be paid to the surrounding manual procedures. Almost all computer systems imply the existence of particular manual procedures, and often are completely dependent upon them. In my own project, experience has shown that the development of such supporting procedures can take as much as 50 per cent of the total development effort on a new project. In particular, procedures inside pathology laboratories vary enormously—far more so than the computer systems.

How much will it cost to transfer?

Many factors come into the answer to this broad question. One such factor—how far does it meet my requirements—has already been dealt with. Other factors which might well be considered are:

Programming language There is an increasing tendency to write in high (or, at least, higher) level languages than before. Such languages as Fortran, PL1 and Cobol, which have international currency, are now very common. However, implementations of compilers for such languages can and do differ from manufacturer to manufacturer. Fortunately, this tendency is becoming less, and high-level languages with their closer approach to English and hence readier comprehensibility will become still more common.

Low-level languages (such as user codes) are usually particular to an individual model of computer, or at best a range. They are usually designed to make particularly good use of features found within the architecture of that computer (and hence unlikely to be found within other ranges). In general, the appearance of a program listing in a usercode is a great deal less self-explanatory than the same program in a high-level language.

Operating system interface Every application program makes frequent use of an operating system supplied and written by the manufacturer. Effectively, an operating system schedules the various resources of the machine, handling requests for facilities made by the application program or programs running at that time. If terminals are involved, it will be the operating system which polls them (looking for requests for service) and maintains a record of whether they are in conversation and at what point.

The particular interface discussed here can be briefly defined as how far the application program has to perform its own work and how far the operating system will help it. An example might be the need to access a particular record in a particular type of file (say, indexed and sequential). A sophisticated operating system might, if given a record key, open the file, pass through various levels of index, check overflow blocks if the key was not present in the prime location, and mark the record (once found) as 'in use' to prevent simultaneous updating. At the other extreme, a basic operating system might do none of those functions and so they would have to be written as part of an application program, with a considerable expenditure of effort.

As a generalisation the more powerful the computer the more complex the operating system. Where a range of machines is offered, manufacturers operating systems are usually upwards compatible (i.e. systems which will run on computers low in the range will run on higher computers). The converse, however, is by no means necessarily true.

File size It is a truism that no two establishments are alike. If clinical laboratories are considered as an example, then volumes of requests and tests performed vary (as may also total repertoire of tests and test : request ratios). It should be then borne in mind that the file sizes quoted by the host site will be a function of the prevailing level of activity of that location. Other factors may influence the file sizes; an obvious one affecting pathology laboratories is how often the file is culled to remove records no longer current (i.e. for discharged inpatients). The requirement for immediate access to non-current records may vary widely from place to place.

Documentation The presence of adequate documentation is self-evidently helpful. Apart from the possibility that in the case of transfer between dissimilar computers it will form the only large scale map of the system, it will serve in the future as probably the only record of what the designer intended. He or she will have moved on, if the current job stay of slightly under two years in computing is maintained.

However, more arrant nonsense has been talked about documentation standards than most topics. It cannot be stressed too strongly that it is the existence of a set of adequate, and maintained, standards that is important rather than whose name is on the standards. I hold to the heretical view that no particular advantage is served by imposing the same set of standards globally. The vast majority of computing staff joining my project have been unaware of either the format or the content of the standards. However, none of them has found the slightest difficulty in reading existing specifications (as, indeed, no one capable of reading an index should with any set of standards) and writing to the new standards has usually meant a trivial amount of work with the standards manual. This experience confirms my own, having written to five different sets of standards (so far) without any trouble whatsoever.

Documentation of this type is used, in my experience, for three main purposes:

1. to obtain a good general concept of what a particular conversation/program does and how it fits in the system/suite.

2. to obtain a detailed knowledge of a particular conversation/program, either for comprehension, or to enable modification(s), enhancements or rewrites to be successfully implemented.

3. a further situation–the emergency repair–is also served by the same means. This case, which is perhaps the most urgent, is the acid test of how good the standards are.

Obviously, everyone will and can have their own ideas about standards. Here are some of ours

a. *General description:* This section gives a non-technical description of the conversation/program. It is of the order of three or four typed pages. As changes are made it is updated. This section is by itself sufficient for the person seeking only a general view of the conversation/program.

b. *Schedule of modifications, enhancements and rewrites:* Dated enquiries showing when the change was implemented, its nature and who programmed it.

c. *Macro flowchart:* This gives the interrelationship between the various modules of coding which make up the total unit. It gives a general understanding in a semitechnical way of how the modules fit together.

d. *Micro flowchart set:* For each module there is a detailed flowchart at a level of detail sufficient to permit direct programming. There is also with it a narrative describing the function of each module. A list of entrance and exit points shows which other modules use them. Various other points are noted such as core requirements, files used, and standard subroutines used.

e. *Test data:* Test data and results are held separately from the specification but not held after the conversation/program appears to be running satisfactorily.

f. *Input documents/formats:* Formats and 'typical' filled in documents or screens are shown.

g. *Output documents:* Formats and 'typical' output tabulations or screens are shown.

CONCLUSION

The experience with the three large industrial groups, all seeking transferability by one means or another, is reasonably supported by National Health Service attempts to achieve the same end nearly a decade later.

The following appear to be the main points:

a. The attempt to design, from the start, a new system to be transferable is difficult and becomes vastly more difficult as the number of parties involved rises.

b. The demand for transferred systems is likely to arise by local initiative rather than nationally. One simple and basic reason for this is that a finished product, running successfully, is more attractive than the prospect of participating in the design of a transferable system (no matter how meaningful the role).

c. The cost of transfer has to be considered closely.

d. The genuinely transferable system will tend to be written in a high-level language.

e. The transfer will either be to a similar but not necessarily identical machine or be run on the host machine with remote terminals (in the case of a real-time system).

f. The documentation will be adequate and exact for the purposes of assessment for transferability and subsequent maintenance and improvement of the system.

g. In my opinion, only pressure of demand leading to successful transfers will now add to our knowledge about the technical skills required to achieve success. It is an area where probably enough generalised thinking has taken place, and where the next step is to specify from the hard lessons of practical experience rather than theory what criteria should be followed towards the goal of easy transferability.

Part 2

MICROPROCESSORS

SOME REMARKS ON MICROCOMPUTERS AND MICROPROCESSOR SYSTEMS

G Jack Lipovski
Department of Electrical Engineering
University of Texas, Austin, Texas

INTRODUCTION

The Microcomputer is computer technology's superstar of the 70s. On one hand, the microcomputer looks like any other computer, except for scale. There is nothing new. On the other hand, the microcomputer is the universal large scale integrated (LSI) circuit, and is the prima donna of the LSI revolution in two major ways. The microcomputer has been delineated as the application of LSI to general purpose processing (1). The microcomputer, the fractional horsepower computer, has also become a mere component in electronic systems of all kinds. These two facets have incredible ramifications.

We who ride aboard this vehicle, computer technology, become complacent about our own rate of acceleration. As Bill Davidow of Intel Corporation points out, (2) in thirty short years from Univac I to the microcomputer, computers have become six orders of magnitude faster, and four orders of magnitude cheaper and more reliable. Accelerated technologies like this have created positive major changes in our life style and social behaviour, maintains Davidow, but have also created wars of increasing devastation. We who ride aboard computer technology have something to ponder, to extract the good from our impressive technology. Even so, the computer is still so feeble compared to the human brain that we can look forward to even more impressive accomplishments in the future.

This article will focus on microcomputers, what they are and what we can and cannot expect of them, to the extent that such analysis and prediction are possible now in this rapidly changing technology. To the extent that microcomputers look like other computers, this article does not attempt to explain their principles of operation. The reader is assumed to have a working knowledge of computer architecture. Rather, some contrasting features of microcomputers, compared with conventional machines, will be analysed.

The Microcomputer, an LSI device

What is a microcomputer? This question has to be answered in terms of computer architecture and of integrated circuit technology. While we assume the reader has a working knowledge of the former, we find it expedient to introduce some terminology about integrated circuits to introduce the microcomputer.

A large scale integrated (LSI) *die* or *chip* is a thin sheet of a semiductor, usually silicon, in the order of 1 cm^2, on which about 1,000 gates have been built up. Since many LSI dice can be fabricated in parallel by photographic techniques with very little human intervention, they are very cheap. Although individual dice can be bought, most are mounted for convenience of handling and for structural protection in plastic or ceramic dual in-line packages (DIPs) whose size is between 0.3 x 0.7 in and 1 x 2 in. Connections to the dice are made by pins on the package. We often informally refer to either dice or DIP packages as 'chips'. Generally, some tens of DIP packages are mounted on an epoxy glass board (printed circuit board), in the order of 10 x 10 in, on which copper conductors have been photographically laid out to connect various pins. Printed circuit boards are generally plugged into sockets much like pages are bound in a book and the assemblage is mounted in a case.

Though the following terms are rather freely used, common practice has been sufficiently consistent that we will venture a definition. A *microprocessor* is an LSI chip or a small set of LSI chips, which embodies the controller and data operator (arithmetic logic unit), of a conventional von Neumann computer. A *microcromputer* is a chip, printed circuit or collection of printed circuits in a case that realises a complete von Neumann computer (controller, data operator, memory, and the digital electronic part of the input/output module) in which at least the controller and data operator are on an LSI chip or a small set of LSI chips. A microcomputer with its associated mechanical input/output devices is usually called a *microcomputer system.* The term, microprogramming, does not refer to programming microcomputers, contrary to a rather popular misconception. It refers to programming the controller to interpret the instruction set of any computer, whether it is a large computer or a microcomputer.

The microcomputer is not really an architectural concept. Indeed, at least four minicomputers (the PDP-8, PDP-11, Nova and TI990) have been redesigned as microcomputers having the same architecture (i.e. instruction set and I/O connection rules). Some optimists claim that someday the controller and data operator of large computers will be put on a very large LSI chip; we will look into this possibility later. But we would venture to call such a computer a microcomputer.

Ironically, our superstar was born of a broken marriage. At the dawn of this decade, desk calculators of considerable complexity were being put on LSI chips. Why not a primitive computer, then? First, Fairchild produced a hybrid between a calculator and a microcomputer, called the PPS-25, and Intel followed with their hybrid called the 4004. While these are classified as microcomputers today, they did not have the general purpose Princeton class (i.e. von Neumann) architecture that is properly a computer, and not a calculator, architecture. Datapoint Corporation, a leading and innovative manufacturer of intelligent terminals based in San Antonio, Texas, contracted with Intel and Texas Instruments to produce a microprocessor of the von Neumann architecture on an LSI chip. Based on their being the largest buyer of MOS devices, they strongly encouraged reluctant Intel and Texas Instruments to embark on this ambitious venture. The design failed to meet specifications, being an order of magnitude too slow. The contract was broken. Evidently, the father of the microcomputer was burned by the venture. Once burned, twice cautious. Even to this day, Datapoint declines to use microprocessors in its products, relying instead on conventional medium scale integrated (MSI) circuitry. A recession followed, which led the large and diversified Texas Instruments Corporation to cancel its microprocessor effort. Intel was not so well insulated from the recession, manufacturing only a good memory chip to back its investments up. It had to market the microprocessor to recover its design costs. Thus, the first microprocessor, the Intel 8008, was made available. It was primitive and clumsy to program. Some say it set computer architecture back ten years. It required the support of about thirty MSI packages. A microcomputer took almost as much printed circuit board space as a conventional computer of the same power. However, it was successfully manufactured and sold. It was a triumph of the LSI revolution; it proved that a microprocessor was a reality and it created a brand new market where there was none before. In the competitive environment of the electronics industry this first entry into the market gave Intel a significant edge and their current Intel 8080 microprocessor, successor to the 8008, is the largest selling in the world!

The venture was so successful that virtually every semiconductor manufacturer has joined in pursuit of the exploding microcomputer market. Recently (1976), some 44 microcomputers have been catalogued by Ryoichi Mori (3). From such a table, the reader may observe that most current microcomputers have 4 bit, 8 bit or 16 bit word lengths, 4k to 64k words of addressable memory, and about 70 instructions in the instruction set, and can add two numbers in about 10 microseconds.

The cost of a popular microprocessor such as the Intel 8080 is in the order of $10, while less popular more sophisticated microprocessors cost upwards of $600. Microcomputers on a single printed circuit board with about 512 words of primary memory, range about $250, while multiple printed circuit board microcomputers housed in an attractive case and having about 8k words of memory are being sold for about $1000.

According to a survey conducted by Mini-micro Systems magazine on the basis of 10,000 returned questionnaires (4), about two-thirds of the half million microprocessors planned for production in 1977 are 8 bit word length machines, about one third are 4 bit machines, and one percent are sixteen bit machines. One third of the microcomputers to be installed in 1977 will be Intel 8080's (5). The Rockwell PPS-4 (6), a 4 bit hybrid between an electronic calculator and a computer, particularly useful in electronic point-of-sale (cash register) terminals, is second in popularity, accounting for about one fifth of the market. The Motorola M6800 (7) places third in popularity, sharing 15 per cent of the market, followed by the Advanced Micro Devices AM-2901 (8) and the Fairchild F8 (9), each having about 5 per cent of the market.

The reader might observe that currently we think of a microcomputer as an 8 bit word length machine with a modest amount of primary memory, with a reasonably useful instruction set, and with speeds one or two orders of magnitude slower than large machines, but with costs three to four orders of magnitude less than large machines. One can put a $10 microprocessor in almost any moderately priced electronic system. One can buy a small but workable microcomputer for half the cost of a colour television, and a useful (though still modest) computer system for less than a car. That, in short, is what all the excitement is about.

Microcomputer Instruction Sets

This section aims to show that in general the microcomputer is a rather powerful computing instrument. The characteristics of 8 bit computers are sketched, both in support of this claim and as a reference for the next section.

The instruction in a typical 8 bit word length microcomputer is commonly stored in one, two, or three consecutive words. The first word generally stores the operation code and addressing mode codes while the second and third words generally store immediate operands, direct addresses, or constants used in addressing modes. The fields of the operation code are usually encoded in an intricate pattern to fit all instruction and address mode codes of a machine into 8 bits.

Most microcomputers utilise the single accumulator organisation, with the addition of 6-64 scratch pad registers and possibly some index registers. Although some limited operations (e.g. increment) can be performed in the scratch pad registers, almost all arithmetic-logit instructions are executed on the data in the accumulator. The programmer can expect AND, OR, ADD, SUBTRACT, COMPARE and ADD/SUBTRACT with carry, as needed for multiple precision arithmetic in this small machine. Two's complement, unsigned binary and decimal arithmetic are directly supported with instructions, but floating point arithmetic and multiplication and division have to be implemented by means of subroutines or macros. Some monadic instructions like CLEAR, INCREMENT, and COMPLEMENT can be executed on the accumulator. MOVE instructions can move words between scratch pad registers, and between the accumulator, scratch pad registers, and primary memory. Further discussion on the MOVE instructions will be considered when addressing is scrutinised. Edit instructions generally include left and right rotate (circular shift of the accumulator and carry together, useful for multiple precision shifts) and two or three other shift instructions. The remaining shifts and edit operations are done by short subroutines or macroinstruction sequences.

Microcomputers usually use condition codes such as carry, zero, negative, and overflow, set by arithmetic-logic instructions, and tested by conditional branch instructions. Conditional branch instructions are generally two word instructions in which, if the condition is true, the second word is added as a two's complement offset to the program counter. A jump instruction is included with a 16 bit address to jump anywhere in primary memory, beyond the range of the branch instructions. Subroutine jump instructions generally save the return address in a first-in last-out stack which is generally stored in primary memory. Software and I/O interrupts are implemented in simple fashion, getting the new values of the program counter from a word in primary memory, and simple I/O instructions may exist just to move data between the accumulator and I/O device registers. Alternatively, I/O registers are addressed as if part of primary memory (memory mapped I/O) and the load and store instructions are used for input and output.

Early microcomputers relied entirely on pointer addressing (index addressing without an offset) to memorise or recall data in memory and immediate addressing to generate constants. Direct addressing, page zero addressing and limited index addressing (where either the displacement or the index register are limited to 8 bits) were added. Indirect addressing has been avoided because the equivalent function can be had by reloading the pointer register. Relative addressing has been left out, except in branch instructions, because most applications would have the program in read-only memory, where local data cannot be both memorised and recalled. (Note that constants can be stored as immediate operands, so they would not be accessed by relative addressing.)

Microcomputers are generally as good as any machine regarding logic and branching instructions, and nearly as good regarding arithmetic and edit instructions except floating point and multiplication or division. Their short word width make them exceptionally suitable for character string manipulation, and their multiple precision instructions make them adequate for typically 24-32 bit commercial or 40-64 bit scientific numeric processing. Their biggest problem has been addressing modes. However, the problem has been recognised and is being corrected in newer machines.

We believe that microcomputers could be as easy to program as large machines, even though no existing products can even pretend to that claim. When that goal is achieved, though, microcomputers can process data at, say, a tenth the speed but at a thousandth the cost of larger machines. A substantial part of data processing will be suitable for microcomputer processing.

DIRECTIONS FOR FURTHER DEVELOPMENT OF MICROCOMPUTERS

Microcomputers are still very young, so that predictions about future developments are hard to come by and not very accurate. Nevertheless, some observations about what is going on and some predictions about where we should be headed may offer the reader some interesting, and hopefully useful perspectives. Therefore, we propose, in the next two subsections, what we believe would be useful directions for the development of instruction sets and of organisational techniques.

Directions for Further Instruction Set Development

The instruction set is the interface between the hardware and software in any computer. Microcomputer instruction sets face some unusual problems because of the limited capability of a small LSI chip and of the wide range of applications. They affect the ease of programming, and therefore the cost of programming, in a significant way.

We often talk of generations of computers, or more particularly, generations of architectures. Our classification of generations of microcomputer architectures is the following.

The first generation, including the Intel 8008, Fairchild F8, and Rockwell PPS-4, concentrated on simply fitting the bare essentials on the chip. Characteristically, addressing modes were sacrificed because it was felt that one mode was adequate, and addressing logic was 16 bits wide and used up a lot of chip 'real-estate'. The designers had to be superb device technologists. A company found its best flip-flop designer, and told him to design a microcomputer. Of course, the intersection between the set of good device technologists and the set of good computer programmers is practically empty. These machines gave their programmers migraine headaches.

The second generation, including the 8080, M6800, and to some extent the TMS 9900 (10) and Z-80 (11), concentrated on making microcomputers as efficient to program as minicomputers. Characteristically, more addressing modes and index registers were added. The designers were quite often hired away from minicomputer manufacturers; they brought with them the time-proven techniques and some of the hang-ups of minicomputer designs.

It is hard to draw a boundary, but the third generation of machines begins in some sense with the TMS 9900 and Z-80. It will concentrate on exploiting the peculiar characteristics of microcomputers. In some senses, this tendency appears in the second generations. For instance, the observation that memory and microprocessor speeds are comparable lead to putting registers in memory in the TMS 9900. This is not acceptable in a minicomputer, where registers in the ALU are ten times faster than magnetic core memory. However, we believe that this generation will exhibit significantly different directions in the development of instruction sets, as we now delineate.

A hard principle to swallow is that there is no perfect universal microcomputer architecture. Any advertisement to this effect should be ignored. Currently, since most microcomputers are stripped down minicomputers, they are quite comparable in speed and storage density for most benchmarks. They may differ by a factor of two or three but they are not orders of magnitude different. The PPS4 is, however, quite different. It fits into a class of calculator-computer machines which is tuned to the point-of-sale application. To an extent, we believe that third generation microcomputers will become more specialised to take advantage of the developing markets. We see two rather distinct classes and at least one additional class that might develop.

The first class includes high volume and/or simple process machines used to replace hardware. As Adam Osborne points out (12), these are dominated by the cost of replication, not by either hardware or software design costs. Consider, for example, a system that sells 100,000 units at $10 each. If the hardware design cost is $10,000 and the software cost is $30,000, their combined design cost is still insignificant compared to the cost of replicating each copy. It is more important to shave $1 off the replication cost than to decrease software cost by half. Paramount to the replication cost is the organisation and realisation of the machine, and the architecture has to live within and support good realisation techniques. The F-8 is an example, where the minimum practical system consists of two chips (the microprocessor and a special memory chip) and the architecture is restricted to operating within the capabilities of the special memory chip. Moreover, such applications rarely require more than 1k words of program memory, and are hand programmed to fit snugly inside the fewest possible memory chips. Techniques like stacks to support recursive or reentrant programming get in the way. Such applications concentrate mostly on moving data to and from I/O registers, intercepting tables of control parameters, delaying actions by means of wait loops and following decision trees. Behold the mighty computer, able to invert a matrix in a single bound, pacing back and forth in a wait loop most of its life. One should discard almost everything except the increment and the lowly move instruction but make the move instruction as powerful as possible. A conditional move instruction, which is a move that is completed only if a condition code had been set, is very useful in evaluating decisions based on McCarthy formalisms rather than nested IF-THEN-ELSE expressions. These humble but useful instructions should appear in a third generation machine designed for simple process control.

A second distinct class of machines should develop around software intensive applications. Intelligent terminals are especially software intensive. As Raymond Yeh of the University of Texas maintains, software costs outstripped hardware costs in large machines by a factor of three in 1975, and are projected to outstrip them by nine in 1985. This ratio would likely be even more striking in microcomputers. The key problem in software-intensive applications is to reduce software cost, even at the expense of adding more logic. This problem is more like the traditional problem in designing large machines, and techniques from larger machine designs can be borrowed to solve it.

Stack machines, in which all arithmetic is executed on stacks in primary memory, are particularly suited to microcomputers in software-intensive applications, especially since the multiple precision problem has been solved for stacks (13). Even large stack machines such as the Burroughs 5000 series store each instruction in an 8 bit 'syllable', several syllables to a word, because stack instructions need no address bits and are thus naturally short. More significantly, software-intensive machines will be programmed in high level languages, and stacks directly implement compiler-produced intermediate level code without 'glitches' such as those requiring the saving and restoring of registers. A compiler can be written for a stack machine for significantly less cost than for other machines because it doesn't need to program for such possible glitches. As demonstrated by the success of the Hewlett-Packard calculator line that uses stacks, a lot of assembler language programmers would prefer a stack machine because it is cleaner and most programs take fewer steps. However, the stack machine philosophy is something of a religion. Opponents claim, with some justification, that stack machines are slower than single or multiple accumulator machines. True believers like Bill McKeeman claim it is better to do the right thing slowly than to do the wrong thing fast. We believe that software-intensive applications should use stack machines.

This second class of machines should also incorporate memory protection, protected instructions, and capabilities, to prevent software bugs from propagating from module to module in large software systems. While memory costs have dramatically fallen, so have microprocessor costs, so that a $10 microprocessor might be supported today by 32k words of memory, at a cost of $1000. Virtual memory would serve to support such a cheap microcomputer with much less real memory, but such that the programmer would enjoy the luxury of programming in a large address space. Virtual memory hardware for microcomputers has been shown to be very simple (14). Lest someone call this sheer wishful thinking, such a microcomputer incorporating multiple precision stacks, protection capabilities, and virtual memory was designed in 1973 and is patented (15). Although the chip size was actually smaller than that of the contemporary 8080, this microcomputer was never manufactured because marketing analysis failed to show a sufficient demand for such a machine. But perhaps in the near future, there may be more awareness of the needs of software-intensive applications and a third generation of microcomputers will be designed for them.

Bearing in mind the low incremental cost of adding microprocessors to existing microcomputers, a third class of machines should emerge that can exploit multicomputing. The most popular instruction advocated to exploit multicomputing is the test-and-set instruction. I/O techniques for packet switching between microcomputers have also been suggested for linking multicomputer systems. While the test and set instruction and packet switching techniques would be of value, we believe that 'vari-structure' techniques to link byte-wide microprocessors 'at run time' into larger word-width processors and vector orientated processors (16) may be more useful. Data flow techniques (17) might be useful at the subroutine level, too. However, microcomputer designers cannot refer to common minicomputer techniques, since they do not apply on the scale that is attractive for multimicrocomputers, and while universities have proposed techniques that do look attractive, they have lacked the funds to test out these ideas.

Several practices of second generation microcomputers should disappear in third generation architectures. At least we hope so. The following technique obviates the need for two practices that are borrowed from minicomputers. Consider that two microprocessors, A and B, can be connected through the same busses to memory and I/O, such that one is running while the other is halted. Most microprocessors have this ability to be halted in order to use direct memory access I/O.

If both A and B are the same type, such as an 8080, then A can run the main programs while B is halted, but an interrupt can turn B on and halt A. In effect, the registers in A are saved while A is halted until B is halted and the main program is resumed. There is no need to duplicate the registers in the CPU as in the Z-80, or to put registers in memory as in the 9900, to effect saving and restoring registers, because it is so easy to put two simpler microprocessors without these features into a microcomputer. The latency time using this technique will be even lower, and the system designer can implement more than two microprocessors to get the effect of more than two sets of registers.

If both A and B are different types, such as an 8080 and a 6800, then A can run programs written for the 6800 while B runs those for the 8080. The personal computer hobbyists are doing this now in order to share programs written on different machines. Engineers may also be using this scheme to evaluate different microcomputers without buying complete evaluation kits. It means, moreover, that upward compatible instruction sets are not as necessary to protect the software investment of the user. The user can run the old software using the old computer, but can write new software for a new machine using a completely new instruction set, which does not have to be upward compatible with the old machine. Both the old and the new microprocessors can economically be put in a microcomputer; the old microprocessor can execute the programs written for it while the new microprocessor can execute the new software. In some cases like the upward compatible 8008, 8080, Z-80 series which is patched-up confusing, it is no longer justifiable to patch up further. Neither the microcomputer manufacturer nor the system designer need limit his selection of new machines to include the instruction set of the old machine in order to protect the users software investment.

Directions in Microcomputer Packaging Techniques

Eight bit processors are fairly well designed from the organisational point of view and should continue for a long time to be so organised. The forces of convention and the availability of parts made cheap by high volume production will probably inhibit improvements in organisation based on repartitioning of resources.

What we are very likely to see, however, is continued improvements in device density and larger modules on the same chip. We now discuss one of the intriguing questions: which way will they grow? Will we put 'wider' processors (32 bit wide maxicomputers) on a chip, or more complex 8 bit wide processors on a chip, or more memory on a chip?

To approach this question, we look at LSI technology. Currently, even a microprocessor die costs only $2 to manufacture, and the dual in-line package costs about $1.50, according to an estimate by Owen Williams of Motorola. It costs about another $3 to market and distribute an LSI chip. In large volumes, LSI chips should cost $6.50. A salesman from National Semiconductor pointed out a few years ago that the cost of a device is halved about every two and a half years. Clearly, we are enjoying a period of rapid evolution in device technology. How long it will continue is debatable. Let us be optimistic and consider what will happen if this evolution continues to allow more devices to be on a chip. Will it be possible to put a 'maxicomputer' on a chip?

While the cost of replicating (making copies of) chips goes down, the cost of designing chips does not. According to an off the cuff estimate by Justin Rattner of Intel, it costs $150 per connection to design an LSI chip. (Note that design cost is not proportional to gates but to interconnects.) This cost has been reduced a little bit by using design automation, but decreases in cost due to improvements in design automation have barely been able to keep up with increases in engineer and technician personnel costs (salaries). It will probably cost the same (about $1 million) to design a microprocessor of today's complexity whenever it is done in the next ten years, or longer. A 'maxicomputer' would cost ten or a hundred times as much to design. Now, marketing folklore has it that if you halve the cost of a chip, the number of units sold is increased tenfold. Conversely, if you double the cost of a chip, the number of units sold will be reduced tenfold. How many people need a maxicomputer on a chip if they can also get away with a standard 8 bit word width machine? There may not be enough units sold to amortise the high development cost. Even though a maxicomputer may someday fit on a chip, we may not be able to design it and recover the design cost over the number of units sold.

The argument goes on. Perhaps a computer manufacturer like IBM will put an existing computer on a chip since, as TC Chen of IBM observes, the design cost of copying an existing computer onto a chip is less than the design cost of developing a new computer. Moreover, the advantage of using volumes of existing software on a new 'microcomputer' system is obvious. However, these manufacturers might find it much cheaper to design the computer using MSI or byte-sliced microcomputer systems, even if a maxicomputer could be put on a chip.

What is more likely to occur? Memories are easy to design. Even now in the M6802, for instance, 128 words of random access memory can be put on the microprocessor chip. Given increased LSI technological capability, the advantage of putting 512 words, 1k words, and so on, inside the CPU chip can be had at little design cost. Such microprocessors would reduce the number of chips needed in a system and provide larger less confining software environments for minimal system applications. Equally likely are enhancements to the complexity of the chip. Hardware multiply and divide is often sought to assist in address calculations and special I/O devices that do not consume many pins (such as shift registers or counters) will likely appear within the microcomputer chip as LSI gate densities improve.

Our next discussion involves the direction of development of I/O peripheral chips. Most large companies are currently scrambling to manufacture special I/O chips that interface microprocessors to CRT's, disc memories, or analog to digital converters. However, according to Tom Gunter of Motorola, the manufacturers are constantly being bombarded with requests for variations on their standard I/O chips to such an extent that a more generalised I/O chip is being sought. What might that be? Another microcomputer? Support a microcomputer with another microcomputer? Intel evidently thought so. They have just released their two universal peripheral interface chips, the 8741 and 8041. These chips are almost copies of the 8748 and 8048 microcomputers on a chip, save for slight variations that improve communications between a main microcomputer and this slave microcomputer. Although it is not clear that the instruction set of the main computer is useful solely in handling peripheral functions, the idea of using a programmable interface chip with a stored program seems very attractive, and should lead to new organisational techniques.

One of the main problems in all LSI circuitry is the limitation of pins. Fundamentally, this is due to basic geometry. The number of devices on a chip increases as the area of the chip while the number of pins increases as the perimeter of the chip. Even now, we are severely handicapped by the 40 pins available on a chip. A solution to this problem requires being able to take data from the surface of a chip without going through the perimeter. Light pipes are being developed quickly for telephony and long distance computer communication. Light pipes can accommodate very high data rates, so a lot of data can be sent through one pipe. Though their cost is still too high, they offer some hope in the future. If they pay off, the packaging problem for microcomputers may be radically changed to permit almost infinite access to logic on chips.

APPLICATIONS OF MICROCOMPUTERS

In this section, we survey the kinds of systems in which microprocessors are being used. The first subsection concentrates on the microprocessor as a replacement for logic, and generally on applications where traditional computers have not been able to penetrate cost-effectively. The second subsection analyses the kind of capability that microprocessors can have in more traditional computing systems.

Applications of the Fractional Horsepower Computer

Most microcomputers will find application in other than traditional computing systems. Our intent now is to get some feeling for the classes of applications and the desirable characteristics of microcomputers for these applications. In this subsection, we will survey these applications in four classes: control applications, communications, office automation, and home entertainment, and we will delineate the characteristics of each class.

The largest class of applications is control of both industrial machinery and consumer appliances. We see two categories in this class. They are the logic-timer control and the numerical control categories.

One category is essentially logic-timer control. The controller may translate commands entered by means of switches into combinations of control signals and pulses for some machinery or appliance, and may time and sequence these control pulses. A simple example is a traffic light controller. These systems are characteristically slow. How often can you change the traffic lights? There is no need to have fast microprocessors for these applications. The amount of programming is quite limited, but the advantage of using fewer chips often translates into a need to use unconventional techniques (tricks) to pack the program into a small amount of memory. Whereas these controllers had been implemented with up to hundreds of dollars of relays and timing motors with cams, the microcomputer simultaneously provides a much cheaper solution and a more flexible system that can be redesigned by reprogramming the microcomputer. Moreover, the microcomputer can be used for other tasks. Although some tasks are useless frills that should not be put in, but are put in because 'garbage tends to fill the container' and the microcomputer is a big container, other tasks are very useful and welcome additions. Primarily, the human interface can be better designed around a keyboard and LED display or possibly CRT display. Most microcomputer controlled appliances, such as microwave ovens and television sets, feature keyboard and LED input and output. Secondarily, the microcomputer can collect statistics on the machinery that can be used for preventive maintenance or for modification of the control sequences to avoid excessive wear. Can you imagine a light emitting diode near each joint that is lit by the microcomputer when the joint needs oil? The use of microcomputers in these applications will probably affect more people more significantly than any other use. It will probably revolutionise our way of using machinery and appliances, and significantly alter our life style.

A second category is the numeric control system. Generally, physical measurements are converted into analog voltages, and linear or non-linear control techniques are used. These systems used operational amplifiers and hybrid analog-digital computer techniques. A fairly powerful minicomputer was used to calculate transfer functions. A 16 bit microcomputer, like the TMS 9900 or LSI-11, is an exceptionally good substitute for a minicomputer in these systems because 16 bits has long been recognised as suitable for feedback control systems. A good example of this kind of control system is the fuel injection computer for automobile engines. In order to determine the amount of gas to inject into a cylinder, some tens of parameters (air temperature, engine torque, etc.) are measured, and a cumbersome polynomial having in the order of a hundred multiplications is evaluated on these measurements to determine the quantity of gas to be injected. The problem is so demanding that, according to Ron Temple at Ford Motor Company in 1975, no microcomputer available at that time could fully meet the requirements. However, several luxury cars (1977 Oldsmobile Toronado, Cadillac Seville, etc) are now being sold with microcomputer fuel injection controllers, and virtually all cars will probably have these controllers in the future. The general characteristics of numeric control microcomputers are a need for processing speed and for numeric functions such as floating point multiplications. Designers should be familiar with numerical analysis and the control theory needed to understand the transfer characteristics of motors and sensors. The microcomputer, replacing the minicomputer, brings the cost of these systems down significantly, and thus considerably extends the range of application of these systems into new areas where they had heretofore been too expensive.

We observe a class of applications based on communications, and monitoring via communications. To be sure, high performance computers are needed in telephone exchanges to route messages and telephone calls. Similarly, radio communications may demand high performance computers to schedule transmissions and detect and correct errors. However, the microcomputer by reason of its low cost should considerably extend our ability to control communications systems in smaller sites. A telephone handset that can dial and then compute the cost of a long distance call is such an example. Automatic communications, or monitoring, may become more widespread. A series of monitors on a river can acquire environmental data and 'telephone' this information to a central recording station. Or a health care system can monitor intensive care patients in a hospital or ambulatory patients in their own homes. Home security systems are also in this class, as well as instrumentation for measuring experiments. Generally, these microcomputer systems are orientated towards communication functions such as negotiating for a connection, sending and answering protocols, and buffering the message until it can be sent. They also often interface through analog voltages to measure physical parameters. Although a few systems will have to be fast, most will be reasonably slow. The available microcomputers are quite adequate for most systems. However, since monitoring systems will probably be unattended, and since they can generate spurious, even dangerous transmissions into a larger network, one of the desirable characteristics of communication systems is reliability, both of hardware and software.

The next major category is office automation. Primarily, the microcomputer can serve a larger range of users in the traditional business system application. A doctor's office can maintain his patients records as well as process his bills and serve as a terminal to aid his searching for information in medical data bases. Perhaps even a gas station can soon afford a computer to maintain records on inventory, to process bills, and interface to data systems. Secondarily, larger businesses may decentralise their computing machinery so that each manager has his own system. Perhaps each secretary will have a microprocessor based typewriter to help her correct spelling errors and to interface to the larger machines. Another large application area fairly related to this area is the intelligent terminal used to edit remotely and submit programs in conventional computer systems or queries for library information retrieval systems. Finally, banking and paying bills may be handled without tangible money through electronic funds transfers. While this is technically feasible today, the sociological aspects of computerised banking are very complex, and this application of computers is moving more slowly than first projections indicated. Related to this, the point-of-sale terminal or 'computerised cash register' is, however, widely accepted. One aspect of such office systems is the need for efficient and secure communication. One does not want 'big brother' or unscrupulous associates reading his private information. Another is the need to manage locally reasonably large amounts of data. Floppy discs, magnetic bubble memories or charged couple devices as secondary memories are generally used with and controlled by the microcomputer. Finally, least understood but most important, these systems must be geared to be used by the 'unwashed layman'. It is not reasonable to make all secretaries and clerks into computer programmers, and most people could not even be forced to think in computeresque procedures and formats. The software and interface hardware has to be human engineered. This takes a lot of design effort and wastes a lot of memory and time in the computer. But efficiency in the computer must be sacrificed for user convenience. The microcomputer can be used for good or evil. As we noted in the introduction, we can subject ourselves to inhuman treatment by this Frankenstein, or use this superstar to enrich and fulfil our lives.

A final class of systems is the home system orientated to entertainment. Two categories appear to be emerging. One is the games system and the other is the personal computer.

In a games system, the user interfaces through 'joysticks' and keyboards to play games on a television screen. Electronic games currently use special purpose LSI chips. One popular chip can play six games. Microcomputers can be used so that games, stored on audio cassettes or removable ROMs, can be entered and run. The user does not get bored with a few games, but can buy a new one each month. This category is deceptively simple, but its importance to the computer industry can be seen from this, that the largest single order of microcomputers from a major microcomputer manufacturer was by a game manufacturer. Electronic games and computerised dolls, model railroads and toys are expected to invade the home. Basically these will internally resemble logic-timer control systems, but externally they will look more like toys and as little as possible like computers.

Personal computing systems, on the other hand, will develop along the lines of scaled-down conventional computer systems. Conventional software packages for self-assemblers, text editors and high level language interpreters may appear in read-only memories, while provision for 'microperipherals' will be designed into these systems.

From this subsection, the reader should observe the diversity of applications of microcomputer systems. Judging from the range of potential applications of these systems, one could become very excited and optimistic about them. Even if not all of them develop, the impact of development of just a few will be significant.

The application of LSI to general purpose processing

In this section, we observe that LSI technology changes the ground-rules for all computing systems. Some possible directions for development of LSI based general purpose computers are now outlined.

The primary motivation for using microprocessors in computers is cost. If the hardware cost is divided into three components—processor, memory and I/O—then the second and third costs can remain the same while the first cost is decimated by converting from large processors to microprocessors. Primarily, this gives rise to the inexpensive personal computer, just mentioned in the last section, and to the use of byte-sliced microprocessors to build up the processors of minicomputers and large machines. Secondarily, since microcomputers are so inexpensive, the opportunity exists to use large numbers of them in a computing system. Other ways are the use of parallel architectures for the main processor itself. This question of how to use multiple microprocessors is the central question of Computer Architecture research today.

Flynn (18) categorised parallel computers according to the number of instruction streams (single or multiple) run concurrently in the computer and the number of data streams (single or multiple) processed concurrently. The observations made in his paper are summarised below. A SISD (single instruction, single data stream) organisation is like a standard computer. A SIMD (single instruction, multiple data stream) organisation operates with a single controller, with single instructions, and with several separate but identical data operators with different data in them, operated by the single controller. Finally, MIMD (multiple instruction, multiple data stream) organisations have several separate but identical controllers with different instructions and several separate but identical data operators with different data in them. (The distributed function computer is a close cousin to, and is sometimes classed with, the MIMD organisation. It is covered later.)

The SIMD organisation can be used for APL-like vector-vector operations on the data. Moreover, some of the data operators can be 'turned off', based on tests for data values in them, so that only the others operate on their data. It is therefore possible to execute complex problems, such as are found in aircraft collision avoidance systems, in such SIMD machines. The SIMD organisation has the advantage that only one copy of the program is stored. Its execution governs many operations on data in parallel. The SIMD has the disadvantage that some of the data operators may be disabled most of the time if data operators are 'turned off' as a result of a test on data in them.

One of the key questions that appears in SIMD and other cellular organisations is the communication between data operators necessary to solve such problems as partial differential equations. A weather modelling problem requires, say, the solution of a partial differential equation on a two-dimensional grid of data points. If each data point (i, j) is stored in a data operator, then the solution of the equation using relaxation techniques requires that the data operator storing point (i, j) communicate with that storing point (i + 1, j), (i - 1, j), (i, j + 1), and (i, j - 1). The obvious solution to such a problem is to arrange the data operators in a rectangular array physically, so that each operator communicates with its four neighbours. This is basically the organisation of the SOLOMON and ILLIAC IV computers (19), except that the rectangular array is first folded into a cylinder and then folded again into a torus (doughnut). However, it is possible to store a whole column of data points in one data operator so that the communication paths between data operators form a ring. This organisation is almost as powerful as the array and is considerably simpler. Similar connection problems exist for SIMD organisations to transpose, multiply, or invert matrixes, or to recognise two-dimensional patterns.

A significant SIMD organisation is the associative memory. Whereas random access memories locate words by means of an address (e.g. read word 3) and sequential access memories locate words by relative location (e.g.read the next lower word), associative memories locate words by content (e.g. read any word beginning with the pattern ABC). And associative memory is essentially built with logic in the memory. At least enough logic exists for each word in memory to compare the given comparand word (e.g. ABC) with the word. An associative processor uses this kind of an associative memory as its primary memory. Each word and its associated logic is a processing element. Since a single instruction normally issues the comparand, and all processing elements operate in parallel, most such associative processors are SIMD machines.

The associative processor is really quite different from the von Neumann computer. Whereas the latter is based on the add operation, the former is based on the compare operation. Just as von Neumann computers can compare by subtracting and testing for zero, the associative processor can add by comparison. Addition can be done by a process analogous to table lookup. Since all memory words are capable of executing the search operation, and thus the table lookup procedure, in parallel, the associative processor is capable of highly parallel addition. Associative processors are sometimes chosen for SIMD architectures because the word and its comparison logic form the simplest processing element that is useful. They are used even for numeric operations. However, the availability of more conventional data operator integrated circuits for microcomputers may make such SIMD organisations that use them more cost-effective than SIMD organisations based on associative processors. They are also used for non-numeric processes like data management system operations. Non-numeric processing will be considered in more detail near the end of this section.

The MIMD organisation can be implemented so that all data operators share the same primary memory, and this is their only primary memory (multiprocessors), or each data operator has its own primary memory separate from all the others (multicomputers). In yet a third organisation, each data operator has some local primary memory, as in a multicomputer, but shares some common memory, as in a multiprocessor.

In multiprocessors, either the common primary memory can only read or write one word at a time, or it is decomposed into submodules, each of which can read or write one word at a time. In the first case, such multiprocessors invariably have memory conflicts wherein several processors are idle, waiting to get a word from memory. In the second case, an expensive switch is used to connect the various modules in memory to various processors. Thus multiprocessors tend to suffer from memory conflicts or from requiring a complex switch.

In a multicomputer, one of the problems arises from the fact that a program may have to be stored several times in each of the primary memories associated with the data operators. Finally, in a mixed multiprocessor/multicomputer, one of the problems is that of deciding which parts of the program or data should be in the local primary memory and which parts should be in global memory. However, this mixed computer organisation seems to be the most promising.

One problem common to all MIMD computers is that of sharing information and resources. The scheduling of problems on processors, and the reliable communication between processors to avoid lock-up and missed signals, requires carefully designed hardware and software. Nevertheless, MIMD organisations offer some exciting possibilities to execute massive problems using microprocessors.

The distributed function structure has many forms that are implemented in centralised or decentralised systems. These are often MIMD organisations. However, the hardware is sometimes specialised in each processor. Except for problems dealing with the communication of signals over long distances and over slow lines, centralised and decentralised distributed function systems are quite similar architecturally. Distributed function organisations are primarily useful where the set of processors together can solve the problems that will be met, and where the problem can be decomposed into smaller problems for which each processor can be specially adapted. Then each processor can be smaller than one that is required to solve the problems all by itself.

One form of distributed function architecture has long been used. The I/O channel of some computer systems is itself a second, highly specialised processor that shares the processing task with the computer's data operator and controller. Architectures have been suggested in which special function processors handle compilation of programs, operating system functions, searching functions and so on. In fact, architectures have been suggested in which special functions of the compiler are implemented in different processors.

A special case of distributed function organisation is the pipeline. Suppose a process P can be decomposed into a process P_1 followed by a process P_2. If both processes are done on the same processor, then the processor will be occupied for the sum of the times, $t_1 + t_2$, it takes to execute P_1 and P_2, respectively. On the other hand, two processors can be connected so that the first one executes process P_1 while the second executes P_2. The output of the first processor is fed to the second processor when P_1 is complete. P_1 is now free to start a new problem while P_2 is working on the old problem. This is called a (series) pipeline. The throughput is increased because new problems can be started every max (t_1, t_2) time unit, instead of $t_1 + t_2$ time units as in the case of a single processor. Note, however, that the complete problem still takes $t_1 + t_2$ time units to complete. Pipelines increase the throughput but do not decrease the elapsed time to execute the total problem.

Chen (20) observed, however, that pipelining does not provide more throughput than an equivalent parallel organisation. If two (MIMD) processors each execute P_1 and P_2, then a new problem can be executed every $(t_1 + t_2)/2$ time units on the average. The pipeline can do no better than this because $\max(t_1, t_2) \geq (t_1 + t_2)/2$. The pipeline organisation however, does have an advantage over the parallel organisation. If the processors can be specially adapted to execute only their process, they can be made less expensive.

The two generic problems with distributed function organisations are design costs and waste of processing power. There is a hardware cost and a software cost for each system. For each cost there is a design cost, the cost of producing the first copy, and a replication cost, the cost of producing the next copy. For systems with equivalent complexity, the hardware design cost has always been highest, followed by software design costs, followed by hardware replication costs, followed by software replication costs.* If a distributed function organisation requires a separate design of each processor, the hardware design cost may be prohibitive. Thus identical hardware is often used in each processor. Putting different software in each processor results in the specialisation needed for each function. This simplifies each processor compared to MIMD computer processors because it need only store the program and data required for one process, rather than the entire program for all processes. The hardware reproduction cost is lowered because less memory is needed per processor. Even so, it does not reduce the software design time. In fact, it tends to increase the software design time. Distributed function organisations also tend to waste processing power. If one of the special functions is not being used, its corresponding processor is idle. This processor has been paid for, but its power is not able to be used.

The last major structure is the network. Here, complete computers are connected by means of rather sophisticated communication systems. One such network funded by the Advanced Research Projects Agency, the ARPA net, spans the entire continental United States. The purpose of these networks is to share existing computers, some of which are specially constructed like the ILLIAC IV, and others of which have special data bases or programs stored in them.

Computer networks appear to be especially useful for data management. Consider any major industrial corporation with several manufacturing and sales centres. A local computer can be used to manage the data needed to support each centre. Most problems can be evaluated by access to the local computer. Nevertheless, these computers can communicate by means of the network to other computers to answer problems related to the entire corporation.

The use of computer networks to solve scientific, as opposed to business, problems seems to be less advantageous. It is difficult for a user to learn all of the protocols necessary to use another machine. There is a strong tendency for users to use their own machine wherever possible because that machine is more under their control than one at another centre. While computer networks offer great potential, especially for data management, the lack of user acceptance of them may prevent that potential from being realised.

One of the most interesting problems that challenges computer architecture at this time is non-numeric processing. Various structures, SIMD, MIMD, distributed function, and network, or various combinations thereof, may provide a solution to this problem.

A very common use of computer processing is data base management. Large and small businesses, government agencies, and the military maintain and search these data bases. The key operation is the search. However, the amount of data to be stored, maintained, and searched is normally very large, from 10^9 to 10^{15} bits of data. A non-numeric processor is one that is designed primarily to search large data bases.

Large von Neumann computers are used to do this kind of non-numeric processing. Data is organised into records which are stored in secondary or else archival memory. Since a record is usually quite large, the computer can well use a large primary memory to hold at least one record. However,

* We are talking about manufacturers. Editor.

the searching in one record is usually done quickly. Records are constantly being paged into and out of primary memory. A complex channel can well be used for this data traffic. Thus a large computer is used.

A second problem is that a lot of software must be separately maintained for each centre. It is not possible for the entire data base to be searched. The data base is so arranged that, statistically, the most common searches can be executed with little effort. However, since each centre has different requirements for searching, each centre has different statistics, each centre has different arrangements for data and different access algorithms to get data, and therefore each centre has different software support. The total cost of this software is very high. The root of this problem lies in the architecture, a SISD architecture wherein it is not possible for the entire data base to be searched.

One solution to this important problem is to use a SIMD organisation where many processors work in parallel to search the data. The associative memory is particularly useful. However, early associative processors ignored the basic problem that the data base is normally very large. A sequential memory that is associatively accessed may provide a solution for large data bases. Moreover, the conventional associative memory searches words as separate entities (content addressing). There is no relationship between words. If one scrambles the order of words in such an associative memory, it still works. But data are stored in information structures such as relational, hierarchical, or network structures. This translates to a requirement that data are stored in a data structure such that neighbouring words in memory can have an effect on the search for a word (context addressing). The Context Addressed Segment Sequential Memory (CASSM) organisation is one attempt to do context addressing on a large secondary memory disc (21). A 'microprocessor' is put on each track of a fixed head disc to search the data in the track. All tracks (segments) are searched concurrently. It is an attempt to solve the non-numeric problem using a SIMD organisation.

Another solution to the non-numeric problem is to use a network, as was discussed earlier. Also, distributed function architectures might be useful where special purpose processors are designed to do efficient sorting, secondary memory management functions, or other operations required for non-numeric processing. This significant problem is being attacked with considerable interest at this time. Its solution will likely result in a class of computer architectures that might rival the presently used von Neumann architecture in complexity and importance.

We see three types of computing systems that could develop from using multiple microprocessors. Some fifteen years ago there were two classes of computers—business computers featuring decimal arithmetic and I/O processing and scientific computers featuring floating point arithmetic and array addressing. Although this distinction has been blurred in recent machines, it is still perceptible, and will likely sharpen as microcomputers are used. In our opinion, three distinct systems should emerge. Two will be business systems and one will be a scientific system.

The business system with its terminals and disc memories should put intelligence in both. Intelligent terminals and networks of intelligent terminals should evolve and be capable of data entry and interactive processing. Intelligent discs should evolve for data base management processing to process the data where it is at. While either system or both could be added like peripherals to an existing large machine, and this will probably happen in an evolutionary way, the two systems together form a sufficient business data process. The large machine may well become ineffectively used, and then be discarded.

The scientific computing system may well develop around a large switch which interconnects microprocessors, memories, and I/O devices. As opposed to the 'loosely coupled' processors in a network of intelligent terminals, these will be 'tightly coupled' to apply a number of microprocessors to the same task. One technique we've proposed (16) is to have full microcomputers whose carry linkage is accessible so that they can be interconnected, like byte-sliced microprocessors, to form a larger word length machine. The programmer can request a word length that is suitable for his program, and the processors can be configured by means of the switch under control of software at run time. While new capabilities and increased performance are evident, so are a number of serious problems. Foremost among these is the switch used to interconnect processors. The cross-point switch so often proposed is patently too costly and unreliable. Stone's perfect shuffle network (22) and our banyan network (23) are promising switching structures. Even so, the software techniques to coordinate numerous microprocessors are barely understood. Some new techniques like data-flow are very interesting (17). Though the rewards are evident, a great deal of effort is needed to solve the problems of switching and software control to realise that goal.

CONCLUSIONS

What does the future offer? On one hand eminent computer architects like Bob Barton, architect of the B5000, paint the microprocessor as a fad, which will be superseded by LSI sequential machine chips as a replacement for logic. On the other hand, those who have risen in the microprocessor industry, like Justin Rattner of Intel, refer to it as the second computer industry. They imply by such an appellation that it is quite different from the conventional computer industry and is potentially as large, or larger.

The current microprocessor environment, fad or new industry, is developing so explosively that we opt to borrow a quotation from Lewis Carroll's 'Through the Looking Glass' that Bill Davidow used to describe this industry (2):

Alice and the Red Queen had been running very fast and when they stopped, Alice observed, 'Why, I do believe we've been under this tree the whole time! Everything's just as it was!'

'Of course it is,' said the Queen. 'What would you have it?'

'Well, in our *country,' said Alice, still panting a little, 'you'd generally get to somewhere else—if you ran very fast for a long time as we've been doing.'*

'A slow sort of country!' said the Queen. 'Now, here, *you see, it takes all the running* you *can do, to keep in the same place. If you want to get somewhere else, you must run at least twice as fast as that.'*

Whatever else can be said about microcomputers, they most certainly are powerful and fascinating devices and their rate of development is breathtaking.

REFERENCES

1. William, AO and Jelinek, HL, *Computer,* June 1976, Page 34
2. Davidow, WH, *Microprocessors and Education,* Intel Corp., July 1976
3. Mori, R (1976) *Survey Report on Microcomputers,* Microcromputer Committee of JEIDA, Page 135
4. Executive Summary: *1977 Mini-Micro Market Survey* In *Minimicro Systems, 10,* (5), 96
5. Application Note, 8008, 8-bit Parallel Central Processing Unit, Intel, June 1972
6. Soucek, B (1976) *PPS-4 Microprocessor,* In *Microprocessors and Microcomputers,* Wiley Chapter 9
7. *M6800 Microprocessor Programming Manual,* Motorola, Inc., 1975
8. *M2900 TTL Processor Family,* Motorola Inc., 1977
9. Application note, *Fairchild F8 Microprocessor,* Fairchild Semiconductor, 1975
10. *Model 990 Computer and TMS 9900 Microprocessor,* Texas Instruments Inc., Manual No. 943441-9701, July 1976
11. *Z80-CPU and Z80A-CPU Product Specification,* Zilog, 1977
12. Osborne, A *An Introduction to Microcomputers,* Adam Osborne and Associates, Inc., 1975
13. Lipovski, GJ, *Proc. Euromicro Workshop,* Nice, France, June 1975, Page 137
14. Anderson, JA and Lipovski, GJ *Proc. 2nd Annual Symp. Comp. Arch.* January 1975, Page 80
15. Lipovski, GJ *Plural Memory Controller Apparatus,* Patent No 4,016,545
16. Lipovski, GJ (1977) *IEEETC, C-26,* (2), 125
17. Dennis, JB (1973) *Computation Structure Memo, 93,* Project MAC, MIT
18. Flynn, MJ (1972) *IEEETC,* Sept. 1972, Page 948
19. Barnes, GH, Brown, RM, Kato, M, Kuck, DJ, Slotnick, OL and Stokes, RA (1968) *IEEETC, C-17* (8), 746
20. Chen, TC *Comput. Dec.,* January 1971, Page 69
21. Su, SYW and Lipovski, GJ, *Proc. International Conf. Very Large Data Bases,* September 1975, Page 456
22. Stone, HS (1971) *IEEETC, C-20,* Page 153
23. Goke, R and Lipovski, GJ (1973) *Proc. 1st Annual Symp. Comp. Arch.,* Page 21

Proportions of this article were extracted from the articles 'Digital Computer Architecture' Volume 7 and 'Microcomputers' Volume 10 (in press) from Belzer: Encyclopedia of Computer Science and Technology, by courtesy of Marcel Dekker Inc.

AUTOMATION OF AN EMERGENCY LABORATORY WITH MICROPROCESSORS AND A CENTRAL MINICOMPUTER

Hans-Georg Metzler Dipl-Ing
Heiko Pangritz Dr-Ing

INTRODUCTION

This paper reports on the automation of an emergency laboratory in the Behring Hospital, Berlin, working during nights and weekends. The project started early in 1976 and will be finished in autumn 1977.

The instrumentation of the laboratory comprises six measuring devices for quantitative examinations and in addition some test assemblies for qualitative examinations. Altogether over thirty different examinations are offered which may give more than seventy measured values due to some multiple methods. In this emergency laboratory the number of samples per shift is relatively small and incoming samples have at once to be measured, i.e. they cannot be collected to fill chains of analysing automats. Therefore the measuring devices of the laboratory are manually operated instruments.

The main purpose of the project was to computerise the calculation of results from measured values and to produce comprehensive documentation, and quality control. To achieve this, the instruments were equipped with interfaces for on-line data acquisition, supervision of measuring procedures, and data preprocessing.

The measuring devices and their associated applications are as follows:
Analog photometer–substrates and enzymes
Flame photometer–alkaline metals
Bloodgas analyser–bloodgases and pH
Glucose analyser–glucose
Coagulometer–coagulation times
Cell-counter
Centrifuge–haematology
Microscope

From Figures 1a and 1b, which show the device configuration of the laboratory, it becomes obvious that there is a strict separation of the system into a data acquisition part and a data processing part. These two parts are connected via a standardised port. This separation simplifies the development and gives a greater security in case of a computer breakdown. The data acquisition part was realised in the CAMAC instrumentation system and as a central minicomputer we use the Siemens computer S310.

SAMPLE IDENTIFICATION

Each specimen-glass wears an adhesive label printed in the laboratory for direct sample identification. This label shows a machine readable barcode and a line of alphanumeric text. Figure 2 is an example of such an identification label.

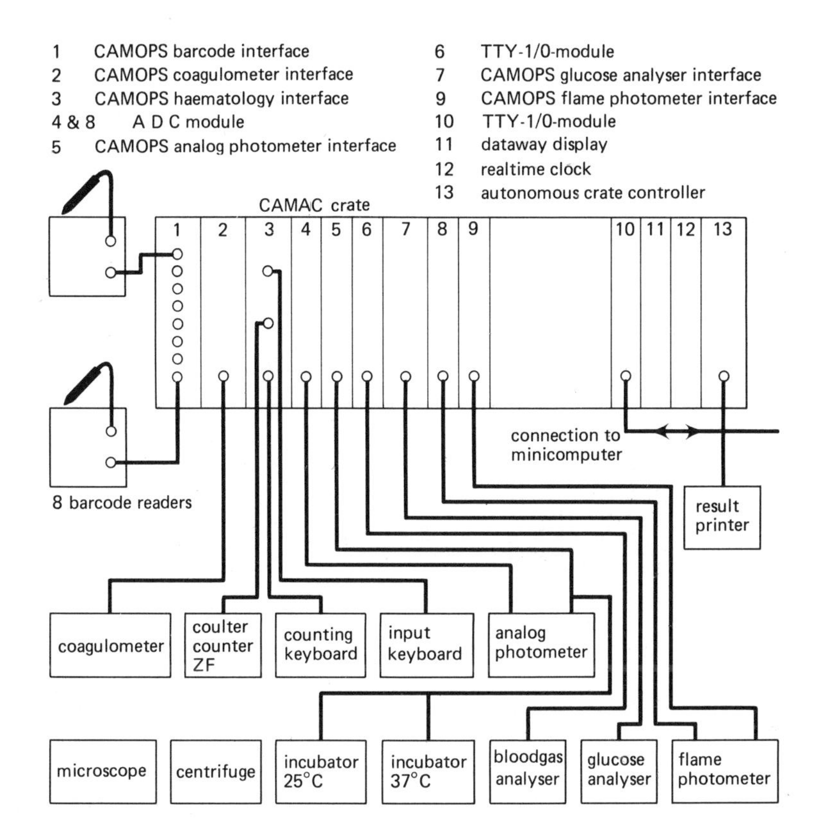

Figure 1a. Data acquisition system

The barcode contains the following elements:

Identification number (I-Number)	6 digits
Method number	3 digits
Check character (start, end, CRC)	4 digits

The alphanumeric text contains:

Patient's name (shortened)	8 characters
Specification of sample material	1 character
Method	5 characters

The I-Number is either the characterisation of the patient of whom the sample originates or it characterises non-patient-specific samples (e.g. quality control samples).

All examinations performed in the laboratory are referred to by a three digit method number. Examinations of the same kind but with different sample material have different method numbers. The individual examinations of collective methods have their own method numbers (e.g. pH in bloodgas analysis).

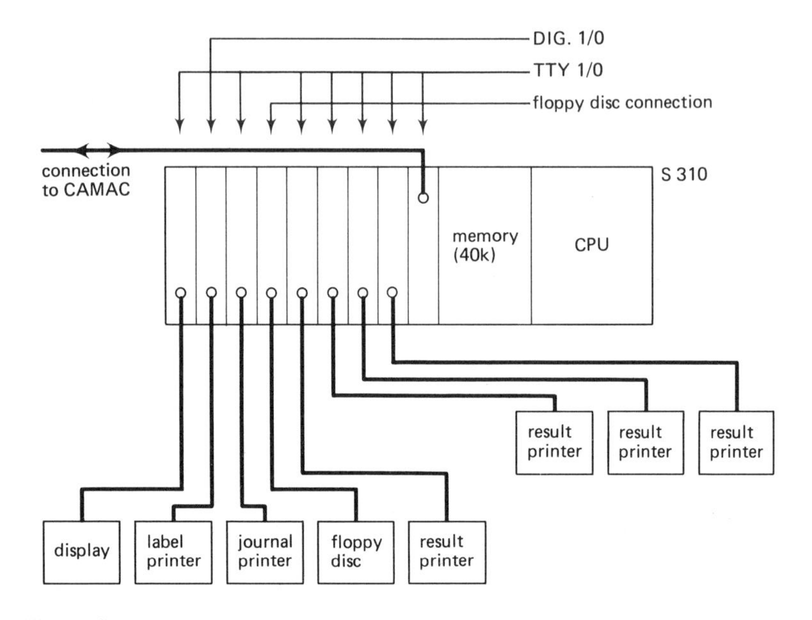

Figure 1b. Data processing system

Figure 2. Sample identification labels

Figure 3 shows the barcoded labels used to identify the tests. Whether a test is qualitative (e.g. urine sediments) or quantitative is indicated by the first barcoded digit.

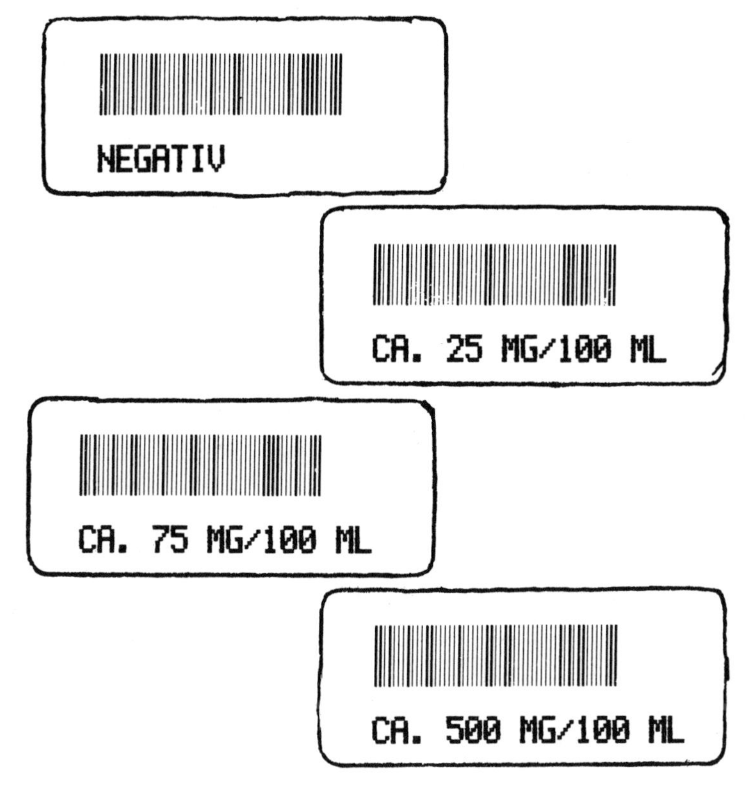

Figure 3. Result labels (total albumin in liquor)

DATA ACQUISITION SYSTEM

The interfaces of the different measuring devices in the laboratory are parts of a CAMAC instrumentation, as shown in Figure 1a. The CAMAC crate itself contains a controlling microcomputer (CAMAC crate controller) and thereby forms an autonomous data acquisition system.

The coupling between the CAMAC system and the central minicomputer for data processing is accomplished by TTY-compatible 20 mA current loops.

If the central minicomputer breaks down or if the transmission line between the two parts of the system is disturbed, the measured data will be printed by a small printer (simplest form of data protection) connected to a serial port of the CAMAC crate controller.

With only one exception, each of the CAMAC interfaces for the laboratory instruments contains a microcomputer for device control, data acquisition, and data preprocessing. The one exception is the blood-gas analyser, which itself contains a microprocessor and which is connected to a simple CAMAC teletype module.

Some of the microprocessor controlled interfaces are described below in greater detail.

CAMAC microcomputer

Each 'intelligent' CAMAC interface consists of a standard microcomputer card and if required, a second card for device specific hardware (line-drivers, optocouplers etc.). Figure 4 is a block diagram of this 'CAMOPS' module (CAMAC-modular-processor-system).

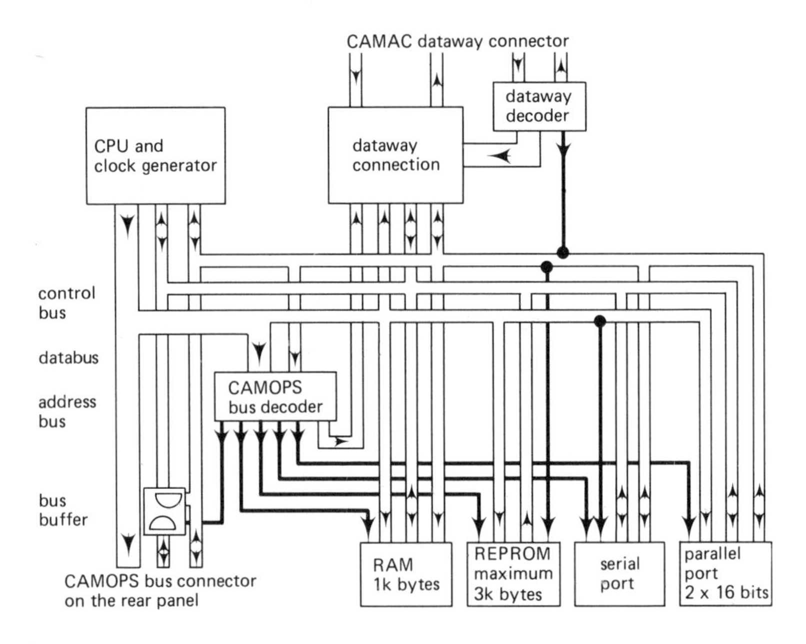

Figure 4. Block diagram of the CAMOPS module

The CAMOPS is based (like the crate controller) on the Motorola microprocessor MC 6800 as a central processing unit. The module contains the following components:

CPU MC 6800
Clock generator (with 'memory not ready' input for adaptation of slow memory chips)
1k static RAM
Maximum 3k u.v. erasable PROM
Serial standard port (TTY or RS232)
Flexible parallel I/O port (TTL compatible 32 bit data and 8 handshake signals)
Asynchronous port for CAMAC dataway connection (128 words FIFOs for each transmission direction).

The two cards of an interface, i.e. the standard microcomputer and the special hardware for the different external instruments are connected via the mentioned standard ports of the microcomputer. This applies to all digital signals. In case of analog signals (here, the two photometers) commercially available CAMAC modules for analog to digital conversion are used, and their data are transferred to the associated CAMOPS module by a servicing program in the CAMAC crate controller. Since the number of laboratory instruments with analog outputs is decreasing, it did not seem worthwhile to design a CAMOPS interface for analog signals.

Barcode-CAMOPS

There are eight barcode reader pens distributed in the laboratory which are not assigned to special measuring devices. These pens are all connected to one CAMOPS by means of simple pulse shaping circuits. Figure 5 is the block diagram of the barcode hardware.

When reading a label a strobe signal is generated at the first light to dark transition. Triggered by this strobe signal the CAMOPS identifies the pen and thereafter only the signals of this pen are analysed until the end of a reading. During this short time possible output signals of the other pens are ignored.

A detailed description of the Beckman Glucose Analyser (BGA) interface is used as an example of the connection of laboratory devices to CAMOPS.

Glucose-analyser-CAMOPS

At the beginning of a new shift the BGA is gauged. Afterwards a quality control measurement is performed. Device-drift supervision tests with known glucose concentration are necessary every hour. With these tests the device is adjusted by the laboratory assistant.

In addition to acquisition and preprocessing of patient's test data the above described procedures are controlled by CAMOPS. Therefore the position of two rotary-switches in the device had to be made available on-line to CAMOPS.

Besides the BCD-data of the device internal digital voltmeter CAMOPS needs the signals 'read' (measuring finished), 'fill' (fill syringe in operation) and 'drain' (measuring chamber empty). CAMOPS sends three signals to the BGA, two for driving display elements, and one for voltmeter statisation. Figure 6 shows the block diagram of the BGA CAMOPS.

The two lamps ('EDP' and 'dilution') show the BGA-status evaluated by CAMOPS. The 'EDP' lamp is on if the BGA interface has received a data block consisting of I- and method-numbers indicating that a barcoded label for a test involving the BGA has been read. This 'EDP' lamp will be switched off after termination of a measurement. It will be flashed periodically if either the gauge or the supervision test is not performed successfully or if more than one hour has passed since the last supervision test.

The dilution lamp comes on if the fill syrine was operated twice (double quantity of enzyme solution in the measuring chamber) so that the next measurement value would be multiplied by two in CAMOPS. The lamp is switched off after operating the drain syringe (emptying of the measuring chamber). It flashes periodically if the measured value exceeds the range of the device ($\gtrsim$ 500 mg/dl).

The CAMOPS program allows for the predilution in urine examinations and the glucose concentration difference between plasma and serum.

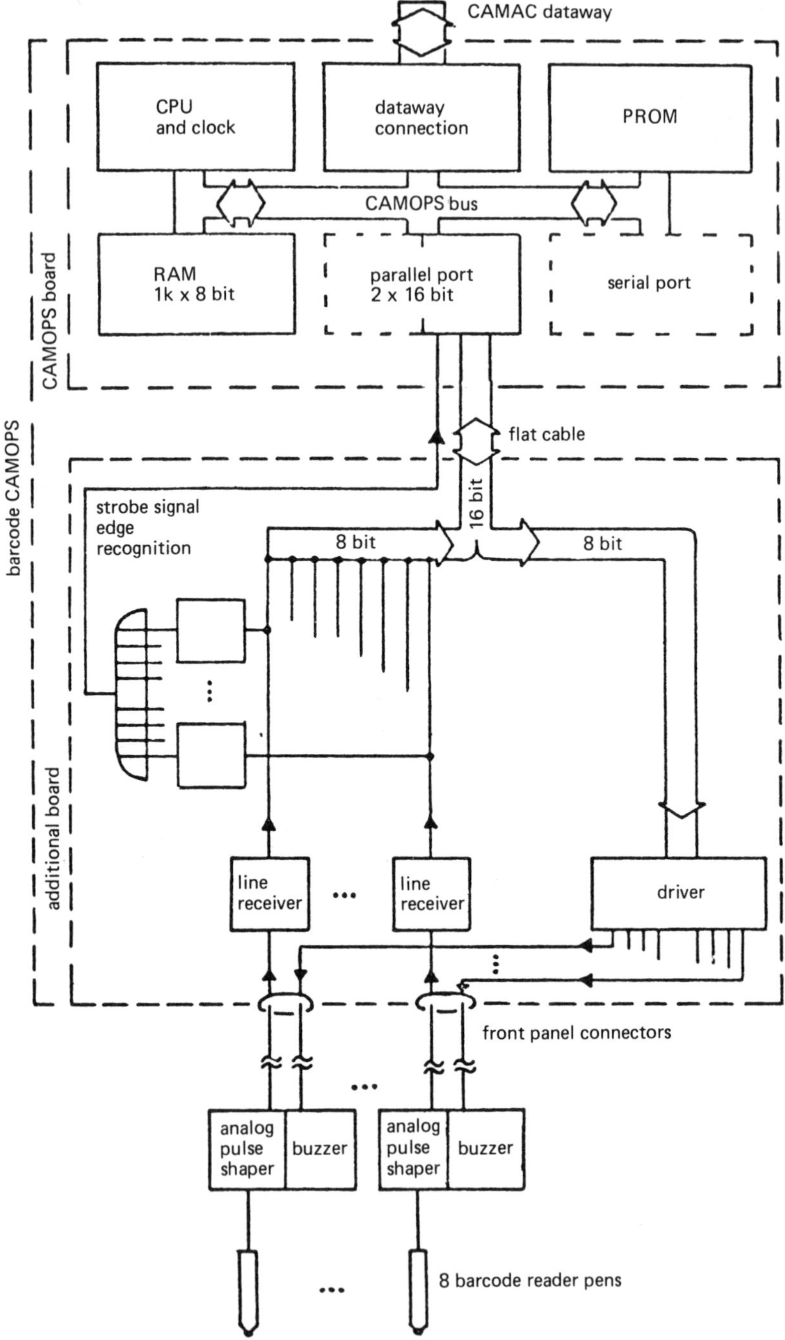

Figure 5. Block diagram of the barcode-CAMOPS

After termination of a measurement a data block is built up which consists of preprocessed measuring data, sample identification number, dilution factor and one byte signalling measurement failures. The block

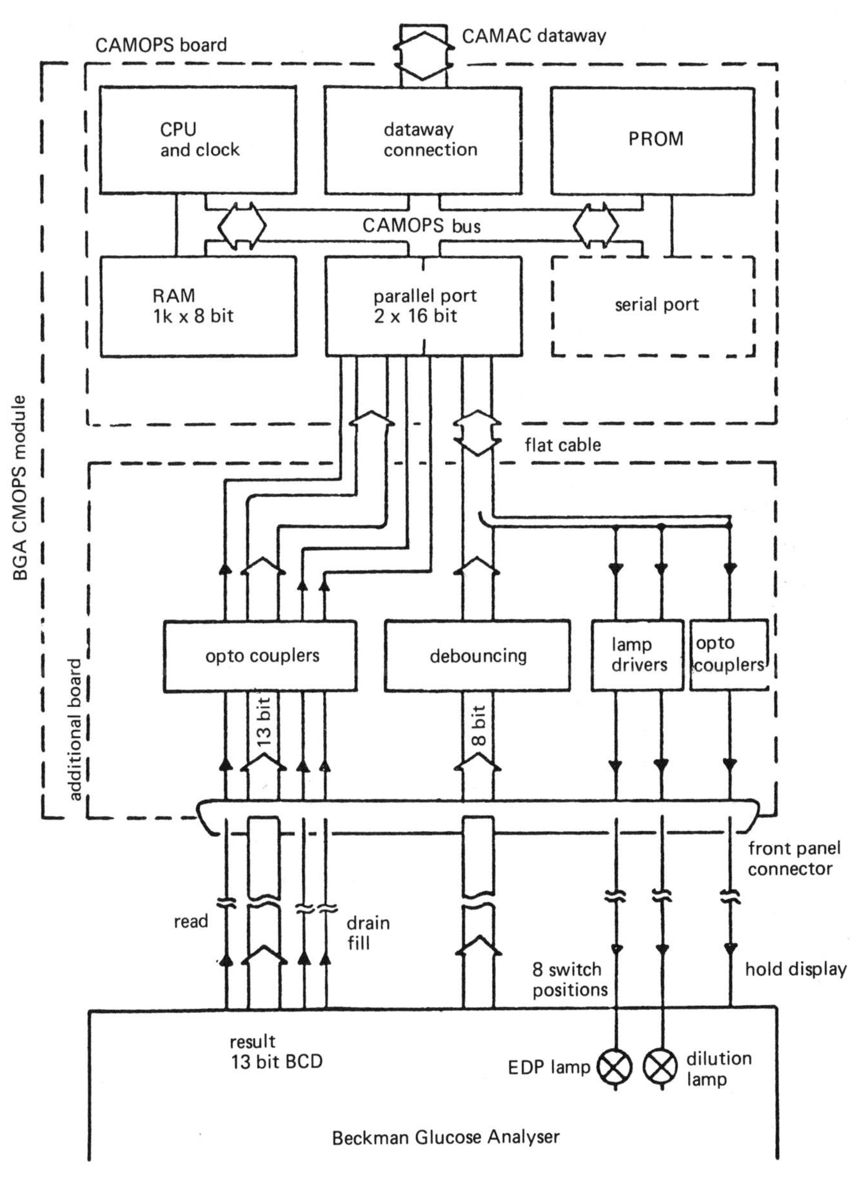

Figure 6. Block diagram of the BGA-CAMOPS

fulfils the GMDS specifications (1). The data are sent to the crate controller which transfers it without further processing to the central minicomputer.

DATA PROCESSING

Hardware

The central minicomputer (Siemens S310) has a memory size of 40k words (16 bit) which is made necessary on account of using the BASIC interpreter (22k).

The following peripheral devices are connected to the computer:
One display with keyboard
Two floppy disc drives
One wheel-type printer (for laboratory journals)
Four result printers (stripe-printers)
One barcode label printer

Three result printers are placed in the wards with the largest number of samples. The fourth strip printer is placed in the laboratory.

One floppy-disc contains patient data and test results. The other contains programs and non-patient-specific files (e.g. text files, print-parameter files). Since one floppy cannot include all programs (120k words) and for avoidance of frequent disc changing there is projected an extension to four floppy disc drives.

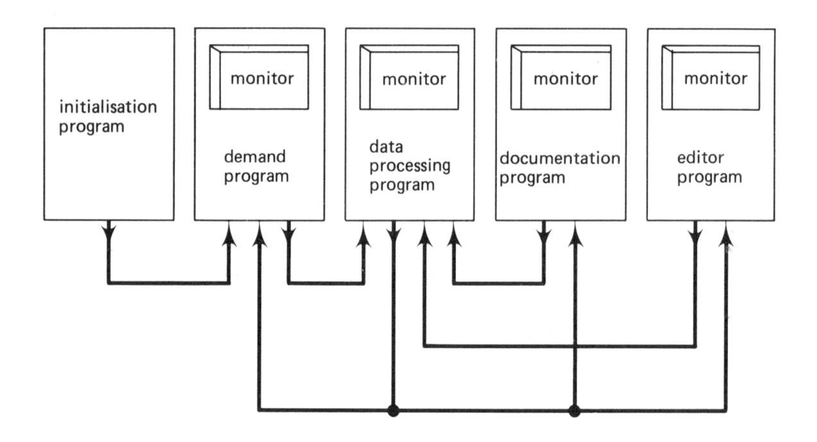

Figure 7. Program segments

Program survey

Some limitations of the BASIC interpreter were overcome by writing a number of assembler routines. It is now possible to use the keyboard in a reactive mode at the same time as other on-line activity to CAMAC is going on and also to have random access to the disc. Since there are only 18k words core memory at the users disposal all programs cannot run at the same time in the core memory. Therefore a segmentation of the

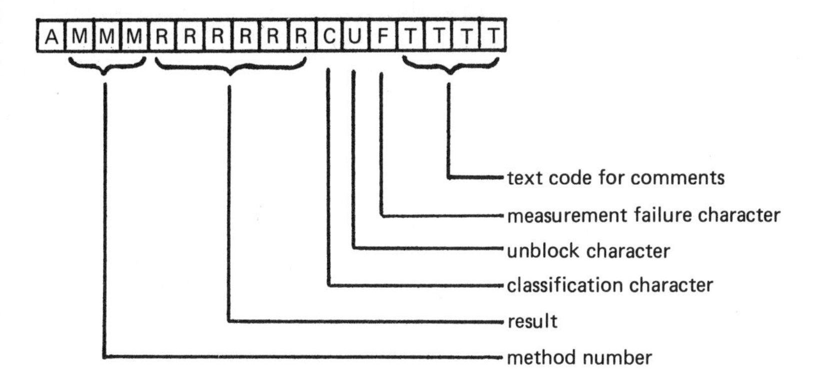

Figure 8. Result field in the patient's document

program system was realised. The arrows in Figure 7 show the possible changes between the segments. All segments except the initialisation segment include the same monitor. This monitor supervises the communication link to the CAMAC system, inputs from the keyboard and the supply of barcode printer.

The initialisation program generates time parameters and a definite start status. This program runs only when the system is set up. The demand program handles the test requests which are entered via the keyboard by the laboratory assistant. The data received by this dialogue produce the barcoded labels for the different examinations. The program keeps patient's documents in a patient's stock file. The patient's document is opened by the first request entry for a patient and is extended by further requests. For each requested test a result field in the patient's document is established (Figure 8).

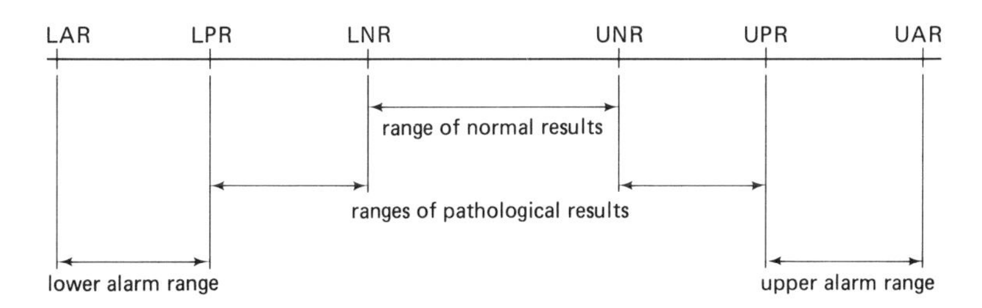

Figure 9. Classification scheme

The measured data processing programs convert the data delivered by the CAMAC system into relevant medical results. The patient documents result field also includes patient specific parameters (e.g. urine quantity and collection time). Reagent dependent parameters are stored in a file for method constants (e.g. reference curve for fibrinogen-concentration).

The results are classified by the schema shown in Figure 9. The method specific limits are part of the method constants file.

A data block consisting of result, patient's data, classification text and eventually a failure comment is printed by a result printer in the laboratory and is transferred to the display. There the laboratory assistant may unblock the results after inspection. After unblocking, the result and the patient's personal data are printed by a result printer in the requesting ward.

By the documentation programs, laboratory journals and daily results lists for the different methods are printed out. Analyses and documentation of quality control results are also made by these programs.

The laboratory personnel may change data of the method constants file (e.g. after changing a reagent) with a special editor program.

REFERENCE

1. Gesellschaft fur Medizinische Dokumentation, Informatik und Statistik (1975) Hardware-Schnittstellen fur den On-line-Anschluss von Geraten im klinisch-chemischen Labor an Datenverarbeitungsanlagen (deutsch und englisch)

AN EVALUATION OF THE PEAK MONITOR OF THE TECHNICON SMAC SYSTEM

I W Percy-Robb, D Simpson, R H Taylor and L G Whitby
Department of Clinical Chemistry,
The Royal Infirmary, Edinburgh, Scotland UK

INTRODUCTION

In 1957, Skeggs (1) described a single channel automatic chemical analyser, a working prototype of the continuous flow system which was soon to be widely adopted for routine use in clinical chemistry laboratories. The first Autoanalyzer* system had its output displayed on a strip-chart recorder, and the chart record served two important functions. It provided the operator with a means of visually monitoring the performance of the equipment and, assuming that the performance appeared to be satisfactory, the individual peak heights could be read off and converted into the corresponding concentrations for the substance that was being measured.

With the development of the Simultaneous Multiple Analyser (SMA)* system (2), the continuous analogue output signal that appeared on the chart records of single-channel analysers was replaced by an oscilloscope, to serve as a function monitor, and by a strip-chart recorder that displayed in turn, on a single preprinted and calibrated chart record, the peak heights of the individual analytical channels that together made up the SMA system; concentrations could be directly read off the chart. Regular and frequent observation of the function monitor by the equipment's operator was required in order to ensure satisfactory operation.

Further development of the continuous flow system of analysis led to the introduction of the Sequential Multiple Analysis plus Computer (SMAC) system. In SMAC the multichannel function monitor of the SMA systems was replaced by a dedicated process-control computer, and the operator of a SMAC system relies on this computer to check the quality of performance of each channel. If the computer records a fault in operation, it is then possible to monitor the analogue output of the faulty channel on a two-channel strip-chart recorder, by manually selecting the appropriate recorder settings. The satisfactory operation of the SMAC system depends upon the efficiency of the computer program in checking the quality of the analytical traces, and it is clearly essential that the software should be sensitive enough to detect unacceptable changes in curve shape.

Two SMAC systems were obtained in 1975, for detailed evaluation in UK laboratories, and the programmes for the two evaluations were coordinated. One of the principal tasks of the Edinburgh evaluation

*Technicon Instruments Corporation, Tarrytown, New York 10591, USA

was to investigate the performance of the process-control software in detail. Many other descriptions have been published of computer assisted monitoring and curve-fitting procedures, as applied to the output of continuous flow analysers (e.g. 3-6) but this account differs from earlier descriptions in at least two respects. Firstly, it relates to a system, the Technicon SMAC analyser, which has been introduced into routine service operation into several hundred laboratories, whereas most of the computer assisted monitoring and curve-fitting operations, to which reference has been made, have so far found very limited application other than in their authors' laboratories. Secondly, because of the commercial nature of the SMAC package, this account draws attention to some of the difficulties that can be encountered if the performance of a proprietary system is found to be deficient. Principally for these reasons, this account of certain components of the SMAC evaluation carried out in Edinburgh differs from previous descriptions of SMAC evaluations (7, 8), both of which concentrated on the performance of SMAC as an automatic chemical analyser, but did not deliberately stress the process-control functions.

THE SMAC PEAK MONITOR

Table 1 lists the 20 analyses included in the Edinburgh SMAC configuration. The detector devices on the potassium and sodium channels are ion-selective electrodes, and on the remaining channels flowcells at which the absorbance of the fluid is measured. Readings are taken from the detectors every 240 ms and preliminary processing of these readings eliminates interference from air bubbles; the net effect is that one in every three readings is examined by the peak monitor.

Table 1. Analyses Included in the Technicon SMAC System as used in Edinburgh

albumin	iron
alanine aminotransferase	lactate dehydrogenase
alkaline phosphatase	phosphate
aspartate aminotransferase	potassium
bilirubin (conjugated)	protein
bilirubin (total)	sodium
calcium	total CO_2
cholesterol	triglycerides
creatinine	urate
glucose	urea

The manufacturers provide users of the SMAC system with sufficient information for them to understand the general method of working of the peak monitor. We do not intend to repeat this account, nor are we in a position to give full details of the peak monitor. Instead, the description that follows is intended to provide the basis for understanding the experiments which were designed to test the peak monitor, and to improve its functioning so as to overcome deficiencies that were revealed by the initial tests.

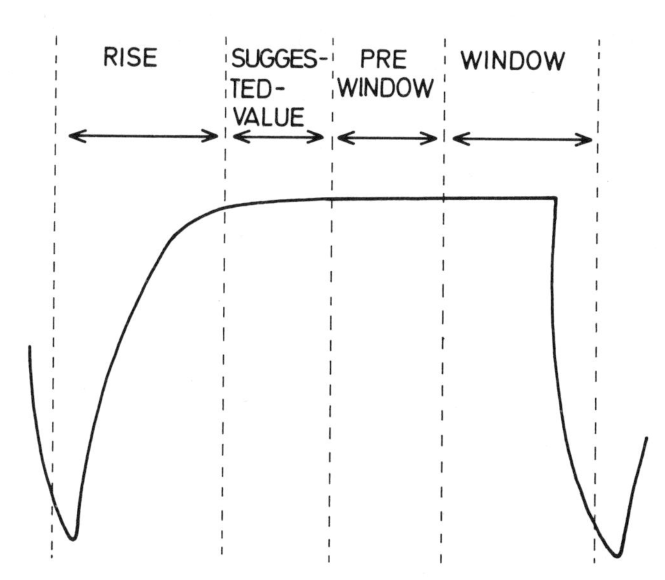

Figure 1. Diagrammatic representation of the recorder trace obtained with the SMAC system, showing the component parts of each curve that are examined by the peak monitor software

The peak monitor checks the quality of each curve in four basic modes, designated as rise segment, suggested value, prewindow and window segments (Figure 1). During the rise segment, a minimum prescribed change in absorbance above the baseline must be achieved, and the suggested value segment provides a frame of reference against which the flatness of the prewindow and window segments may be compared. The parameter which controls the lengths of the suggested value segment, and hence the number of data points recorded during it, has been given the name 'Seg3specs' (segment 3 specifics).

Prewindow starts immediately after the suggested value segment, and lasts for the same time. During the prewindow and suggested value segments, the curves are assessed for noise by comparing the latest reading with the four previous readings. If a curve passes the specified tests of acceptability up to this point, the program proceeds to monitor the window segment.

The functions performed by the peak monitor program in the window segment are

1. to recognise the fall of the curve,
2. to determine the value of the curve immediately before the fall, and
3. to compute the result for acceptable curves.

There are prescribed time limits for each channel within which the fall of the curve must occur for a curve to be acceptable, these limits being defined by the parameter called 'Windowspx' (window specifics). The program also compares the average values recorded for the curve heights during the suggested value, prewindow and window segments, these values needing to be prescribed channel-specific tolerances for a curve to be accepted; the parameter specifying these tolerances is called 'Valuetols' (value tolerances).

Results for acceptable analyses are printed out on the matrix printer. However, if the peak monitor program has detected a fault, either no result is printed out or the figures on the report are accompanied by an asterisk (to indicate that a fault has occurred on that particular channel) or the sample is marked with a rerun signal. The rerun signal indicates that although the peak on a particular channel may have been accepted, faults giving rise to asterisks have been detected on more than 15 per cent of the channels in operation on SMAC at that time, and the appearance of an asterisk implies that no data should be reported for that specimen.

MATERIALS AND METHODS

In the first series of experiments to be described, the SMAC system was operated as a standard production model, with standard software designated by Technicon as SMAC 7.2. All 20 channels were operational, and the SMAC was provided with samples for analysis taken from a single pool of normal human serum. The volume of sample normally taken into this particular SMAC is 420 μl, and the system was presented with a series of duplicate samples, successive pairs of samples having volumes of 2.0, 0.8, 2.0, 0.6, 2.0, 0.5, 2.0, 0.4, 2.0, 0.3, 2.0 and 0.2 ml. Small volumes of sample were used as a readily controlled means of deliberately producing abnormally shaped peaks. The results that were printed out were used to assess the ability of the peak monitor software to detect the presence of abnormal peaks.

In the second series of experiments, the SMAC system was operated with modified Seg3specs, Windowspx and Valuetols parameters, determined following a series of tests with the 'Ticker' software (see later), and the system was again presented with a series of duplicate samples for analysis. The samples were again taken from a single pool of normal human serum, and the volumes in the sample cups were provided in the same sequence as in the first series.

The SMAC system was calibrated with reference materials provided by Technicon, and all reagents were obtained from Technicon except for the Liebermann-Burchardt reagent for the cholesterol channel; this was prepared in the laboratory according to directions from Technicon.

RESULTS

The effect of presenting the SMAC system with samples of decreasing volume is shown in Figure 2a, which is a segment of the recorder trace of the output from the potassium channel when different sample volumes were provided for analysis, as detailed in the legend. With sample volumes of 0.4 ml, the peaks appear normal and there is a clearly visible plateau, but when the sample volume is reduced to 0.3 ml the peaks become narrow and the part of the trace between successive peaks is much wider and deeper than normal.

Figure 3 shows diagrammatically the results printed out for the concentrations of sodium, potassium, total protein and albumin in the samples of the serum pool, as these results relate to the volume of sample provided for analysis. The 'true' value of each constituent is the mean concentration

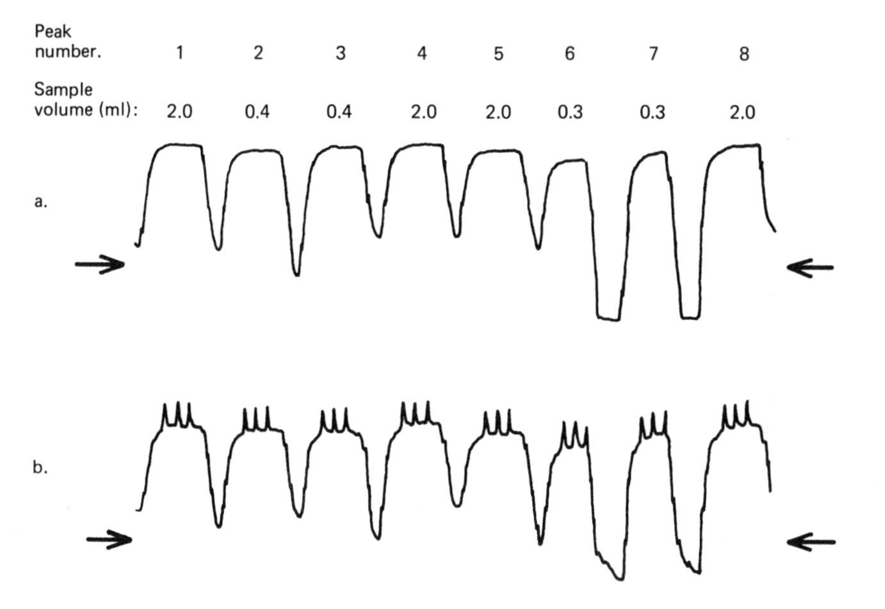

Figures 2a and 2b. Segments of the recorder trace from the potassium channel, showing the peak shapes obtained when different volumes of sample from a single pool of serum were presented for analysis. The arrows indicate the position of the baseline at the start and finish of the group of analyses that included the samples shown on these segments of recorder trace. a. before modification, b. following modification of the hardware and software so as to enable the 'Ticker' program to indicate where the peak monitor has moved from one segment of the curve on to the next

determined by analysing 25 different samples of the serum pool, each sample being of adequate (2.0 ml) volume. Figure 3 compares the results obtained with samples of smaller volume against the standard of accuracy provided by the 'true' value. With samples down to 0.4 ml in volume, Figure 3 shows that there was no significant effect on the accuracy of the results. When the sample size was reduced to 0.3 ml, however, the results were nearly all incorrect (i.e. more than two standard deviations below the true value), not only on the 4 channels illustrated in Figure 3 but on all the other 16 channels. Despite the fact that these results were nearly all incorrect, often seriously, the standard SMAC 7.2 peak monitor program showed a low rate of detection of these incorrect results. It is important to note that many inaccurate results were printed without an asterisk. Also, the rerun signal (intended to reflect the detection of simultaneous

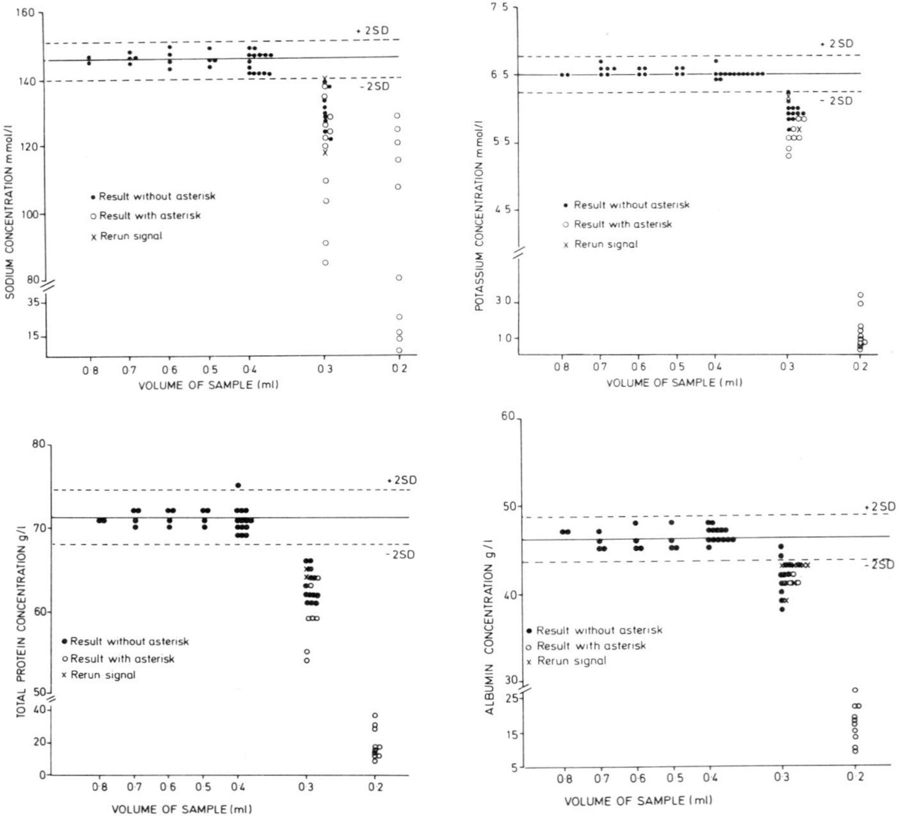

Figure 3. Results obtained on 4 channels (sodium, potassium, total protein and albumin) to illustrate the effects of presenting for analysis different volumes of sample all taken from the same pool of human serum. The continuous horizontal line in each figure represents the 'true' value for the constituent, determined by performing 25 sets of analyses on 2.0 ml samples of the serum pool; the interrupted horizontal lines represent the ±2 standard deviation (SD) limits calculated from these 25 analyses. These results were obtained using the manufacturer's standard software, SMAC 7.2.

faults on at least 4 channels with 20 channel operation) was failing to increase the sensitivity of the peak monitor as a process-control system. When samples of 0.2 ml volume were presented for analysis, a series of grossly inaccurate results were obtained, all of which were rejected by the peak monitor and marked with asterisks.

The manufacturers were informed about the poor performance of the peak monitor, as revealed by experiments such as those illustrated in Figures 2a and 3. In response, Technicon provided us with more information about the process-control function, and modified our SMAC so as to enable us to perform experiments on the programs. The computer memory was extended and the software modified so as to provide the following additional features:

1. The twin-channel recorder was modified so that only one channel was monitored at a time, and the hard copy output on one of the two chart records bore a series of 'ticks', each of which indicated the time at which a change occurred from monitoring one segment to monitoring the next segment (Figure 2b). Because of the appearance of ticks on the chart record, the program was called the 'Ticker' program.

2. The average values for the rise, suggested value, prewindow and window segments of the curve, the mean of the last three of these average values, and coded messages to indicate the nature of any faults detected in each segment of the curve were held in store until the end of a group of analyses and then dumped.

These features enabled detailed examination of the program's function to be carried out, in association with modifications to program parameters. We were provided with information that allowed us to alter Seg3specs, Windowspx and Valuetols, either alone or in combination on one or more channels.

For experimental purposes, as a means of testing the function of the modified peak monitor, the SMAC was again presented with samples of varying size so as to stress the peak monitor in a controlled fashion. The following experiments were performed with the peak monitor parameters:

1. Seg3specs values were increased, so that both the suggested value and prewindow segments became longer. This had the effect of making premature falls of curves, such as occur with narrow peaks (Figure 2a), appear within the prewindow segment thereby leading to rejection of such peaks.

2. Windowspx values were reduced. This also had the effect of making the software more sensitive to the occurrence of premature falls of curves away from plateau values.

3. Valuetols were reduced, so as to make the peak monitor more sensitive to the presence of instability on the plateau portion of each.

Following a number of preliminary experiments, changes in parameters were adopted on the 10 channels for which the shapes of peaks appeared to be best defined. Table 2 shows the values, in octal, for the parameters on these channels as set in the standard SMAC 7.2 software, and the values adopted for use following the modifications evaluated in Edinburgh, designated SMAC 7.2E.

Figure 4 shows data obtained when the experiment described in Figure 3 was repeated, but using a new pool of serum; the data in Figure 3 were obtained with SMAC 7.2 whereas the data in Figure 4 were obtained with SMAC 7.2E. As with the earlier experiments, when samples of 0.4 ml

Table 2. Parameters in the peak monitor software

This table lists the parameter values, in octal, for the software supplied by Technicon as SMAC 7.2 and for the modifications introduced into the Edinburgh SMAC system in order to increase the sensitivity of detection of faulty peaks (SMAC 7.2E). The SMAC 7.2E parameters can be entered by addressing the locations shown.

	Windowspx			*Seg3specs*			*Valuetols*		
Channel	*Location*	*SMAC 7.2*	*SMAC 7.2E*	*Location*	*SMAC 7.2*	*SMAC 7.2E*	*Location*	*SMAC 7.2*	*SMAC 7.2E*
sodium	034355	005	003	034351	005	006	034353	40	30
potassium	034415	005	003	034411	005	006	034413	40	30
total protein	032415	005	003	032411	005	006	032413	150	100
albumin	033455	005	001	033451	005	007	033453	120	100
calcium	032555	005	003	032551	005	006	032553	140	60
urate	032015	005	001	032011	005	007	032013	150	60
bilirubin (total)	033155	005	001	033151	005	007	033153	150	40
phosphate	033515	005	001	033511	003	005	033513	100	70
lactate dehydrogenase (cell 1)	033555	005	001	033551	005	007	033553	130	130
urea	032315	005	001	032311	003	005	032313	100	70

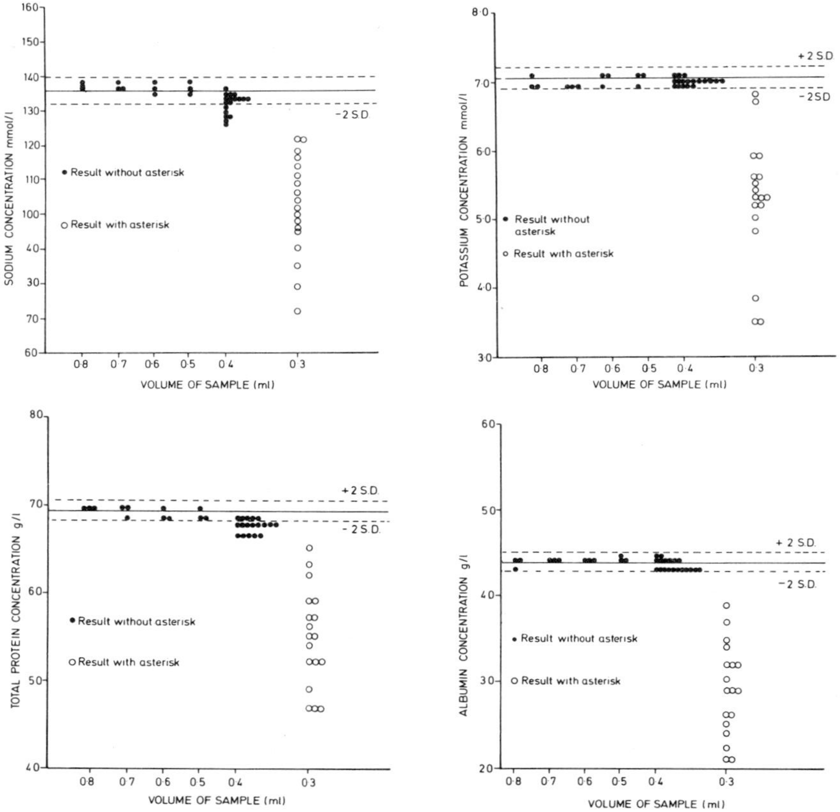

Figure 4. Results obtained on 4 channels (sodium, potassium, total protein and albumin) to illustrate the effects of presenting for analysis different volumes of sample all taken from the same pool of human serum. The continuous horizontal line in each figure represents the 'true' value for the constituent, determined by performing 25 sets of analyses on 2.0 ml samples of the serum pool; the interrupted horizontal lines represent the ±2 standard deviation (SD) limits calculated from these 25 analyses. These results were obtained using the modified parameters for peak monitor software designated as SMAC 7.2E (Table 2)

volume were presented, there was loss of accuracy (quite marked with some of the analyses on the sodium channel) but none of the peaks was rejected. However, when the sample volume was reduced to 0.3 ml, all the peaks were recognised as faulty; not only was every single result printed out with an accompanying asterisk but, because of the frequency of asterisks, all the results were also marked with a rerun signal.

The modified peak monitor parameters (SMAC 7.2E) have been in daily use in this laboratory for over a year. Initially, the SMAC 7.2E parameters had to be entered each day via the keyboard on SMAC, by addressing the locations indicated in Table 2, and this practice has continued following the modifications to all SMAC systems included in the upgrading designated SMAC 7.6. (The peak monitor programs were not modified by Technicon in SMAC 7.6.) Recently, we have been provided with a SMAC 7.6 tape that incorporates the Edinburgh peak monitor parameters, and we no longer need to key these in each day (SMAC 7.6E).

The regular use of SMAC 7.2E (or 7.6E) parameters in the peak monitor has not resulted in an unacceptably high rate of asterisks or rerun signals. Approximately 20 samples each day out of a workload of about 500 samples from patients have had to be reanalysed; this rate of rejection of samples is no higher than when the manufacturer's parameters (SMAC 7.2 or 7.6) were in use. It should be noted that, although most of the samples for analysis in this laboratory consist of specimens of plasma, the amount of downtime attributable to minor blockages of the system averaged only 3.7 min/day, when detailed records were kept over 100 consecutive days' operation.

DISCUSSION

The SMAC system includes software designed to monitor the voltage signals being produced at each output device and to draw the operator's attention to an unsatisfactory set of signals each time a faulty peak is detected. An efficient computer-assisted monitoring system is essential since the SMAC system does not include provision for the simultaneous, direct examination of the voltage output from every channel. The choice of program parameters depends on a balance between:

1. adequate sensitivity in detecting abnormalities in the patterns of voltage output and
2. too great a sensitivity leading to an unacceptable rate of false alarms that an abnormal state exists when, in fact, most patterns of output could be regarded as acceptable. The degree of sensitivity of the SMAC peak monitor in detecting abnormalities should, ideally, be at least as high as would be the case if all voltage outputs were able to be examined on a multichannel display by the operator instead of only two channels being able to have their output displayed.

The manufacturers place great emphasis on their recommendation that the SMAC system should only be used for the analysis of serum. It would seem, however, that this restriction derives more from the fact that the official descriptive literature and the claims for the system's performance were based on data obtained when serum only was being analysed rather than on any positive evidence that plasma is unsuitable for analysis. For many reasons, mainly related to the ease of collection and separation and the lower incidence of haemolysis, plasma is preferable to serum for chemical analysis and many laboratories receive large numbers of specimens of heparinised blood for examination. When we first encountered examples of the peak monitor software failing to detect and reject unsatisfactory peaks, it was suggested that this might be attributed to our use of plasma for analysis.

Partial blockages of the SMAC system might be more likely to occur when plasma samples are analysed than when serum samples only are accepted. As long as specimens are properly centrifuged prior to analysis (9) however, we do not believe that this is the case, and we have experienced very little downtime (average 3.7 min/day) attributable to minor blockages in the tubing. We do not consider that the experiments with inadequate samples, described here, are restricted in their application to SMAC systems where plasma samples are analysed. Irrespective of whether serum or plasma samples are used for analysis, we consider that the SMAC peak monitor should be sensitive to the appearance of abnormally shaped peaks. The experiments described in this paper, however, relate to the analysis of serum samples.

The performance of the manufacturer's peak monitor software has been deliberately tested by presenting samples of pooled serum of inadequate volume for analysis, after preliminary trials of alternative ways of attempting to produce unsatisfactory peaks in a reproducible manner. The alternatives investigated included lifting the sampling probe out of a sample by hand at precisely stated times (e.g. after 16 seconds instead of allowing a full 22 seconds for aspiration of sample), and the application of a pressure clamp to the transmission tubing carrying the sample into the SMAC system. It was considered that samples of inadequate volume might be encountered unpredictably from time to time during regular operation of the SMAC system, following the period of evaluation, the most likely cause of such events being partial blockage somewhere within the system, the blockage often being of only a temporary nature.

When the manufacturer's standard peak monitor software was tested by presenting samples of inadequate volume to the SMAC system, peak shapes were greatly modified (see Figure 2a) but many of these peaks were still accepted and grossly inaccurate results were printed out (see Figure 3). By modifying the peak monitor parameters on 10 channels (see Table 2), the ability of the SMAC system to detect and reject unsatisfactory peaks has been greatly improved (see Figure 4).

The results of these attempts to improve the peak monitor were reported to Technicon approximately one year ago, and the modified parameters are now being tested in a few laboratories in Europe, selected for this purpose by the manufacturers. If our findings are confirmed, and it is accepted that SMAC 7.2E (Table 2) parameters make possible a better standard of performance of the SMAC system, then the modified parameters may be included in a future updating of all SMAC systems. If our findings are not confirmed, or if the use of SMAC 7.2E parameters at the other test sites leads to the rejection of an unacceptably large number of samples, then no action is likely to be taken by the manufacturers, although failure to confirm our findings would suggest that there can be marked differences between the performance of individual SMAC systems, a possibility that would raise interesting questions in relation to quality control of SMAC production.

It is, unfortunately, impossible for others to repeat the many experiments that we performed before deciding to use the parameters listed in Table 2, unless the laboratories were to be provided with the additional information and with the temporary modifications to their SMAC systems that Technicon kindly provided for us. It is, however, possible for others to make use of SMAC 7.2E (or 7.6E) parameters (Table 2) by entering them each day into the appropriate locations during start-up of the SMAC system.

The SMAC system represents one example of the increasing range of laboratory instruments that depend for their operation upon rigid proprietary computer programs. It is understandable why manufacturers are reluctant to issue full details of their software, since this must almost always have involved major financial commitment. The experiments described in this paper, however, show that it is important not to accept the manufacturer's claims for the performance of a computer-controlled piece of equipment until the equipment, and its associated software, have both been independently tested and the claims for quality of performance confirmed.

REFERENCES

1. Skeggs, LT (1957) *Amer. J. Clin. Path., 28,* 311
2. Skeggs, LT Jr, and Hochstrasser, H (1964) *Clin. Chem., 10,* 918
3. Gray, P and Owen, JA (1969) *Clin. Chim. Acta, 24,* 389
4. Griffiths, PD and Carter, NW (1969) *J. Clin. Path., 22,* 609
5. Whitby, LG and Simpson, D (1969) *J. Clin. Path., 22,* supple. *(Coll. Path.) 3,* 107
6. Reekie, D, Marshall, RB and Fleck, A (1973) *Clin. Chim. Acta, 47,* 123
7. Schwartz, MK, Bethune, VG, Fleisher, M, Pennacchia, G, Menendez-Botet, CJ and Lehman, D (1974) *Clin. Chem., 20,* 1062
8. Westgard, JO, Carey, RN, Feldbruegge, DH and Jenkins, LM (1976) *Clin. Chem., 20,* 489
9. Percy-Robb, IW, Simpson, D, Taylor, RH and Whitby, LG (1977) Some aspects of the performance of the Edinburgh SMAC during nine months' diagnostic use. In press.

MAKING BETTER USE OF SCINTILLATION COUNTERS

R P Ekins, MA PhD,
P G Malan, BSc PhD
S B Sufi, BSc
Subdepartment of Molecular Biophysics,
Department of Nuclear Medicine,
The Middlesex Hospital Medical School, London W1N 8AA

Radioisotope scintillation counters, once instruments restricted to physics and nuclear medicine departments, have found increasing use in routine clinical chemistry laboratories during the past few years. This development has arisen primarily in consequence of the explosive impact upon clinical medicine of radioisotopic microanalytical techniques, of which radioimmunoassay (RIA) is the best know example. The fundamental reasons which have led to the widespread use of these analytical techniques in many branches of medicine are two-fold. The first is the unparalleled 'structural specificity' of antibodies and other similar 'binding' reagents (such as specific hormone receptors) *vis-a-vis* substances of physiological and clinical importance. The second is the sensitivity of radioisotopic measurement techniques *per se,* which enables the behaviour in chemical reactions of exceedingly small numbers of 'labelled' molecules to be detected and accurately quantitated. The latter of these attributes, together with the indifference of radioisotope decay processes to environmental influences such as temperature, chemical milieu, etc., imply that radioisotopes represent particularly 'rugged' labels whereby the chemical interactions which form the basis of radioimmunoassay and related assay techniques may be simply monitored.

These two factors ensure that these methods will continue to dominate the microanalytical field in the foreseeable future, and also that they are likely to extend their areas of application to embrace all situations in which relatively complex organic molecules (such as polypeptides and proteins) in low concentration are required to be measured.

The principal challenge to this view stems from the particular efforts that have been made in the past few years to develop analytical methods based on identical principles but relying on other molecular labels such as enzymes, fluorophores, bacteriophages etc., which avoid the technical requirements and hazards associated with radioisotopic techniques. It is not the purpose of this presentation to argue the relative merits of non-radioactive and radioactive labels in microanalysis; suffice it to say that the particular advantages of radioisotopes lie in their simplicity of measurement and their independence from environmental effects. These are likely to guarantee their continued popularity notwithstanding the infinitesimal radiation hazards they present, and the restricted shelf-life that radioactively labelled compounds inevitably possess. For these reasons, it is predictable that radioisotopic techniques *per se* are not likely to be soon displaced from their pre-eminence in situations in which high analytical sensitivity is a paramount requirement.

Nevertheless, one of the major disadvantages associated with radioisotopic analytical techniques arises from the relatively prolonged counting times that must be allocated to the measurement of the isotopic content of assay samples. This arises in consequence both of the small concentration of labelled material normally employed in these techniques, and the random nature of radioactive decay. These factors together imply that a radioactive sample must be counted for sufficient time to accumulate a number of counts that ensures that the statistical variation arising from the random decay process is 'small' when compared with other sources of variation in the system. This requirement lies at the root of the delays resulting from sample counting, and the necessity for the purchase and organisation of numbers of relatively expensive instruments to cope with high sample loads.

This paper essentially concerns the use of microprocessor-controlled isotope counters as an aid in the resolution of this problem. This application does not represent the sole or most important use of computer methods in the RIA or radiomicroanalytical field; we shall be discussing other aspects relating to the analysis of RIA data later in the book. It is nevertheless relatively novel in so far as it illustrates the use of computer techniques in relation to the design and control of RIA procedures, which is an area in which sophisticated data processing facilities can play a unique role. Although dwelling on a particular aspect of this role, the concepts discussed in this paper are generally relevant to all situations in which radioactive samples must be counted.

The fundamental principle upon which the techniques reported here are based is that there is no justification for the counting of radioactive samples for a time longer than is required to increase the precision of the final measurement. The basic rules governing the statistics of radioisotope counting are, of course, well known, and have governed instrument design and usage for the past thirty years. Briefly, they state that the statistical deviation of a sample count is equal to the square root of the total count accumulated; thus, if a total of 10,000 counts is registered, the standard deviation of the count is 100 and the coefficient of variation is 100/10,000 or 1 per cent. Because an error of 1 per cent is generally viewed as being 'small' in relation to other errors incurred in most experimental situations, many workers traditionally count samples for a time sufficient to attain 10,000 counts, although other options, relating both to sample counting times, and to total registered counts, are offered on most radioisotope counting equipment.

That such 'strategies' of radioisotope sample counting may be extremely wasteful of instrument time is revealed in Figure 1, which illustrates the relationship between total sample count and precision, both in the absence of 'experimental' (i.e. sample preparation) and other 'non-counting' errors, and in the presence of experimental errors ranging up to 10 per cent. This figure emphasises that the 'reward' in terms of sample precision in consequence of the accumulation of additional counts may be relatively small when, for example, counts in excess of 1,000 are accumulated in circumstances in which 'experimental errors' amount to 5 per cent or more. This implies that a more intelligent sample counting strategy than is customarily adopted would limit the sample count (and hence sample counting time) to a number related to the pre-existing 'experimental' error associated with individual samples.

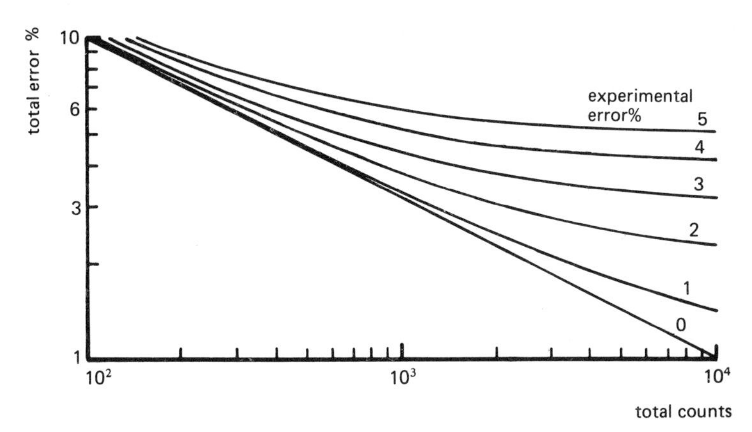

Figure 1. Curves relating overall precision of measurement of samples as a function of total counts in the presence of different levels of sample preparation, or 'experimental' error

In many situations in which radioactive samples are counted, the errors associated with sample preparation are relatively small and constant, and the estimation of the 'experimental' error component in individual samples is quite straightforward. For example, if a radioisotope study were to involve the pipetting of sequential aliquots of blood containing a radioactive substance into sample counting tubes, the sample preparation error would be likely to be roughly constant, irrespective of the radioactive content of the sample, and of the order of between 0.5 and 2 per cent (depending on whether a single pipette were to be used throughout the operation and whether any variations existed in the viscosity of the fluid pipetted). In the case of radioimmunoassay samples, however, the nature of the sample preparation process results in relatively complex 'error' relationships between the amount of radioisotope in the sample, and the experimental error or variation associated with the measurement, quite aside from the statistical error associated with sample counting *per se.* Thus it would not be unusual to observe, in a typical radioimmunoassay, 'experimental' errors ranging, for example, from roughly 2 to 10 per cent, varying as a function of the magnitude of the particular response variable employed in the assay system (e.g. the amount of radioactivity in the sample). In these circumstances, it is clear that the time of counting and the number of counts accumulated for each individual sample in an assay run can be varied without significant loss in precision, depending on the magnitude of the pre-existing experimental error associated with the measurement.

Fortunately, the *pattern* of errors associated with particular (well-controlled) assay procedures is relatively constant, albeit such a pattern necessarily depends upon the nature of the assay system involved. For example, an assay procedure involving the separation of 'free' and 'antibody bound' radioactive moieties using a chemical or immunological precipitation technique without subsequent washing of the precipitate is likely to yield an entirely different form of 'error relationship' from systems in which antibody previously attached to a solid phase support is used and the solid phase linked antibody is subsequently carefully washed to remove all 'free' radioactivity prior to radioassay of the 'antibody bound' material. A characteristic 'error-relationship' can hence be discerned for every assay procedure, relating the error in the response metameter to the magnitude of the response. Thus, if the assay response variable is represented by the amount, or percentage, of the labelled material which is antibody bound, we can, by appropriate statistical analysis, identify the magnitude of the experimental error in the 'amount bound' for any value of this variable and plot a curve, as illustrated in Figure 2, defining the

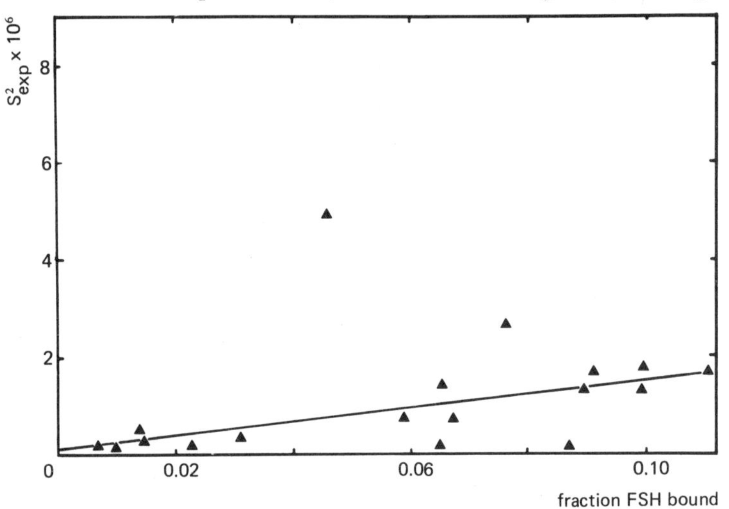

Figure 2. Typical response-error relationship, relating the experimental error in a radioimmunoassay response variable (fraction bound) to the magnitude of the response

error relationship. Such a consistent relationship can be established experimentally either by statistical analysis of a number of preceding runs using the same assay protocol, or alternatively, by setting up a specific assay system using a number of replicates for each chosen value of the response variable. Thus by adopting either, or a combination, of these measures, information can be obtained whereby the 'experimental' error characterising particular values of the response variable can be estimated. Given such information, it is a relatively simple computational matter to arrange that radioactive samples characterised by a high 'experimental error' should be counted for a time yielding a relatively low aggregate sample count (and hence a high counting error), and vice versa.

These concepts underlie the construction of microprocessor controlled scintillation counters built in our own laboratory and subsequently by a commercial manufacturer*. However, it is inappropriate in this presentation to discuss in detail either the electronic design or the microprocessor software involved. Essentially, information is supplied to the micropro-

*L.K.B./Wallac, Turku, Finland

cessor, relating to each batch of radioimmunoassay samples to be counted, defining the experimental error relationship anticipated in the batch. On admission into the counter, an individual sample is initially counted for a few seconds which enables an approximate, and statistically imprecise, estimate of its radioactive content to be made. This estimate, together with other relevant information (e.g. the 'total' counts introduced into each assay tube), enables a rough assessment of the magnitude of the 'response variable' (e.g. fraction 'bound' or fraction 'free') to be made. On the basis of this assessment, the 'experimental error' characteristically associated with the sample is calculated, and a total 'useful' or 'optimal' total sample count, and hence 'statistically useful' sample counting time, deduced. The sample can subsequently either be counted for the time so determined, or repetitive reassessment of the required counting time can be made at intervals (each reassessment being made with improved precision in consequence of the increasing accumulated count) until such time as the elapsed sample count time exceeds the currently required counting time.

Conventional radioisotope counting instruments require that the operator sets the controls so that individual samples are counted either for a preset time or for a preset count. The microprocessor-controlled instruments described in this paper require both information relating to the error relationships characterising individual assay runs (see below) and also a decision by the operator regarding the permitted loss in precision consequent upon sample counting *per se.* With regard to this decision, a number of different options may be offered to the operator. For example, he may adopt a counting strategy whereby the experimental error in the sample is allowed to increase by a given proportion–say 20 per cent, or one fifth. Thus, if an individual sample entering the counter is characterised by a preparation error of 5 per cent, the instrument may be programmed to terminate the count when the *total* or overall error reaches 6 per cent. This implies that the statistical counting error (given by the difference between the total and 'experimental' variances) will be 3.32 per cent and the total sample count required is 909. Likewise, a sample characterised by an 'experimental' or 'preparation' error of 3 per cent will be counted, under the counting instructions described, to a total of 2525 counts.

Paradoxically the characteristic error relationships seen in many radioimmunoassays as currently performed are such that 'experimental' errors increase as the radioactivity in the sample decreases, and this would lead, given the counting strategy described above, to the non-intuitive consequence that samples containing less radioactivity are frequently counted for shorter counting times than those containing higher activity levels.

Other strategies are also possible. For example, counting errors may be constrained to yield an absolute rather than relative increase over and above the sample preparation error. If the permitted increase in error were defined as 1 per cent, then a sample characterised by a preparation error of 5 per cent would be counted for 909 counts as before (yielding a total error of 6 per cent), but a sample with a preparation error of 3 per cent would be counted for such time as to yield an overall error of 4 per cent, i.e. for a total count of 1,428 counts.

The ultimate outcome of these alternative counting strategies are perhaps best portrayed by their effect on the 'precision profile' of the assay, a term which we may use to represent the relationship between the error in the 'analyte' or 'dose' measurement, and the magnitude of the dose. (The precision of the dose measurement may be calculated by dividing the error in the 'response' variable by the corresponding slope of the dose response curve for all values of the 'dose' variable). Figure 3 illustrates the consequences, in a typical assay, of different counting strategies, e.g. the counting of all samples for either a fixed time or to a fixed count, in comparison with an 'intelligent' strategy which specifically relates counting error to sample preparation error. Although the final

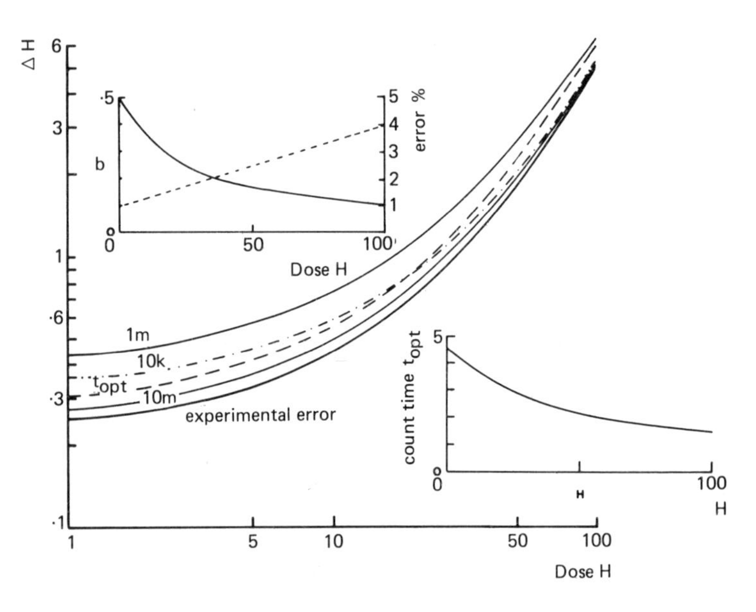

Figure 3. Computed precision profiles in a representative radioimmuno-assay, relating the precision of the dose measurement to its magnitude (ΔH = error in dose measurement). The 'basal' curve is that calculated on the assumption of zero counting error i.e. 'experimental' error only. The other curves are based on sample counting either for preset times (1 and 10 minutes) or a preset count (10,000), or on an 'optimal' micro-processor-controlled strategy based on a 20 per cent increment upon the 'experimental' error. The inserts describe the response curve and experimental error relationship for the assay on which the precision profiles are based (b = fraction bound) and the optimal sample counting times required as a function of sample concentration

precision profiles are discernibly, not not significantly different, it may readily be computed that, in a typical assay embracing samples containing a wide range of concentrations of the analyte, the saving in counting time and in instrument usage consequent upon the 'intelligent' strategy would be considerable.

In practice, it is difficult to quantitate exactly the savings in instrument time arising from 'rational' counter usage as here described. Much depends on the spectrum of concentrations of analyte found in an individual assay run, on the experimental errors characterising the procedure, on the particular counting strategy normally adopted by the assayist and by the parameter values adopted in the particular rational, microprocessor-mediated, counting strategy selected. In practice, we have, in our laboratory, observed savings in counter usage ranging from two to fifteen fold as a result of the change to microprocessor-controlled instruments. A practical illustration of the insignificant loss in overall assay precision in a typical assay resulting from this approach is shown in Figure 4 where the precision of measurement of different concentrations of follicle-stimulating hormone (FSH) are shown following radioassay of samples both in a conventional manner and in a microprocessor-controlled counter. The total saving in instrument time arising in this particular example was seventeen fold.

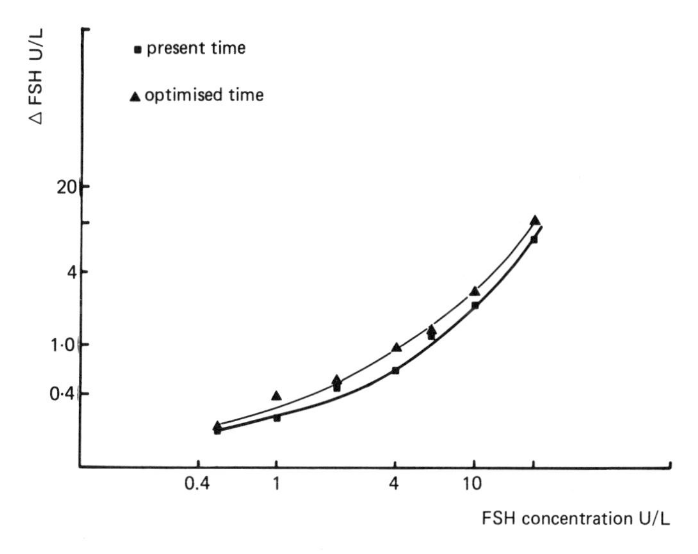

Figure 4. Experimental precision profiles for an FSH assay following a conventional counting strategy, and one based upon microprocessor controlled equipment

Even greater savings in counter time can be achieved by more sophisticated counting strategies than described above. In many clinical situations, for example, not all samples in an assay run require to be estimated with equal or similar precision in order to yield clinically useful data. Samples falling in the centre of the normal range may be estimated with relatively low precision for a diagnosis of abnormality to be definitely excluded. Likewise samples falling unequivocally outside the limits of the normal range may be identified in the face of a relatively low assay precision. Conversely, analytical imprecision attaching to a 'borderline' result must be reduced as far as is reasonably possible in order to maximise diagnostic reliability. This concept is illustrated in relation to the clinical use of thyroid hormone (T4) assays in Figure 5. In this figure two 'borderline' areas are defined: that between normal and low T4 values characteristic of hypothyroidism, and, at the upper end of the range, that between normal and high values associated with thyrotoxicosis. The precision of counting of assay samples can clearly be adjusted so that the further the result lies from either of these borderline values, the less the counting time required to establish an unequivocal result. The upshot of such a strategy is that only those samples in which the result is of critical clinical importance are counted for such time as is required to approach the limits in precision imposed by other experimental factors.

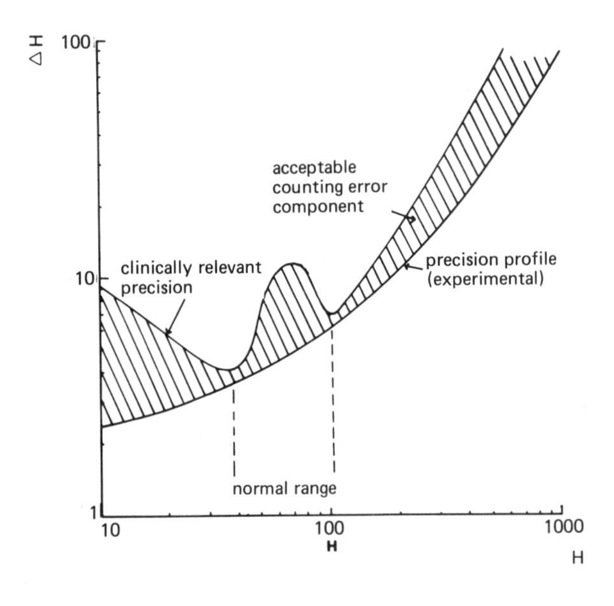

Figure 5. Predefined precision profile defining permissable counting errors at different dose levels

These examples illustrate some of the many different sample counting strategies that can be adopted to ensure that samples are counted only for times which satisfy the precision requirements of the experimenter (or clinician), and which are not overlong in relation to the constraints on sample precision arising from other experimental errors involved in sample preparation. Their implementation essentially depends, in each case, on a microprocessor-mediated feedback control system which, following initial observation of the sample content, computes the time required for the particular sample measurement. Such a counting instrument thus constitutes a truly automated device. These examples also emphasise that, whereas all isotope counting instruments available in the past have been designed to operate in a manner governed solely by counting statistics *per se,* the newer microprocessor-controlled instruments now available will be essentially governed by the precision of the final analytical measurement, taking non-counting error into account.

A fundamental criticism to the approach towards sample counting described in this paper is that a much more sophisticated understanding of the statistical characteristics of assay systems is implicitly demanded of the counter operator. It is worth reminding ourselves, in regard to this question, that most users of radioisotope counting equipment at the present time–even those with a long experience of sample counting in general, and of radioimmunoassay sample counting in particular–are totally oblivious to the ultimate statistical consequences of their choice of a 'preset time' or 'preset count' which represent the two modes of control available on current counting instruments. The approach adopted in this paper makes explicit the fundamental considerations which should ideally dictate counter operation. Nevertheless it is important to provide practical assistance to the counter operator in his choice of settings for counter controls, and to this end, the data processing programs we have developed for use in these instruments analyse assay data with a view to estimating the experimental error parameters upon which 'intelligent' counter operation rely. This means that following a single assay run, the data emerging from the counter is automatically analysed to assess experimental error relationships: which information is, in turn, employed when further assays of the same kind are counted. In short, we are devoting particular attention to the development of data analysis programs for use in these instruments which express, in easily interpretable form, the statistical results upon which future operation of the machine should be based.

The economic benefits of the devices we have described are potentially considerable. The radioimmunoassay load falling on many departments is now considerable and is increasing. Illustrative of very large screening programs, currently under discussion, which would involve the measurement of millions of samples are the detection of hepatitis-B antigen in transfusion blood, the detection of fetal abnormalities by the assay of α-fetoprotein, thyroid-stimulating hormone (TSH) and T4, the detection of tumour associated antigens, etc. Some departments involved in this type of work possess 10 or 20 counting instruments to cope with the sample load and to obviate delays in the derivation of clinically important results. The average price of automatic sample counters is around £10,000, so that the reduction in the number of instruments implied by their more efficient use is likely to lead to large financial savings. One perhaps should mention, in this context, that an alternative γ-counter which has recently appeared on the market tackles the same fundamental sample throughput problem by using an array of 16 detectors, so that 16 samples may be simultaneously counted. This represents an elegant and complementary approach to that which we have described, albeit, the current model is a manual machine, and is not substantially less in cost as compared to the fully automatic, microprocessor-controlled, data analysing instrument based on the approach described in this presentation. To compare these two solutions to increasing sample throughput is out of place in this paper, particularly as each offers logistic advantages which may suit the domestic arrangements of individual laboratories. Nevertheless, future microprocessor-controlled instruments which combine 'intelligent' statistically sophisticated control mechanisms with dual or triple detectors (enabling duplicate or triplicate samples to be simultaneously counted) will undoubtedly enable higher sample throughput rates, and considerably higher daily throughputs (since they are operable for 24 hours a day) to be obtained than by manual multidetector systems.

OPTIMISED AUTOMATION OF CLINICAL INSTRUMENTS (LEUCOCYTE CLASSIFICATION AND BLOODGAS ANALYSIS)

Gerhard K. Megla, PhD, FIEEE
Technical Staff Division Corning Glass Works,
Late Professor, Department of Electrical
Engineering, State University, Raleigh, N.C.

INTRODUCTION

Since the subject of this book is 'Computing in Clinical Laboratories', it might be worthwhile to discuss not only the advantages of computerising clinical instruments for the purpose of automating certain processes, but to show also the limitations of this technology both in practice and in principle. In order to do this I shall discuss two examples of instruments of quite different computer design philosophy which were both recently introduced: a computerised leucocyte classifier and a 'microprocessorised' bloodgas analyser.

Before describing the functions of these two instruments I would like to elaborate briefly on the desirability or necessity of automating any clinical instrument by means of computers. There is hardly any limit to the number of functions of an instrument or system which can be computerised. The main questions to be answered are: does the automation of a particular function make sense; is it needed; is it time and cost effective; does it add real value to the overall performance of the device including features such as accuracy, repeatability, quality control, throughput rates, workflow matching, convenience of operator interaction, ease of service and maintenance. Are, for instance, automatic gearshifts in automobiles desirable or necessary? At present, fewer than 2 per cent of cars in Germany are automatic; on the other hand, more than 97 per cent automatic in the US. The real need of automation in both countries is obviously rather different. One could explain this situation by saying that in the US there is ample and cheap energy available and the user is considering the car merely as a means of transport. In Europe the cost of energy is much higher and, in addition to that, the drivers feel more as professionals do and do not like to have their driving (diagnostic) talents taken away. This is a typical example of incompatible philosophies of automation design because of different economic situations and preferences.

Therefore, the assessment of the added value of automation is different for different kinds of users and this has to be taken into account if one tries to predict the real value of automation. Such predictions are not always right, be it in clinical laboratories or anywhere else for that matter: about 100 years ago the mayor of Chicago predicted by linear extrapolation that in 30 years his city would be so filled up with horse manure that nobody would ever be able to walk the streets. The motorised car was invented and Chicago saved. In 1950 a US trade union predicted that the introduction and wide application of the computer would cause unacceptable unemployment since it would make obsolete the middle class work force. We now know that computerisation does not generate unemployment but rather the governments and the unions are to be blamed for it. Automated clinical instruments designed to replace medical and laboratory staff are missing the real need, which is to free such professionals in clinical laboratories from routine work so that they are able to elevate themselves to a higher level of expertise and professionalism. Most of my life I have been a research director in academic and industrial institutions and know that even up to now some computer scientists would like to computerise almost everything. If we let them do it the result would be the replacement of red tape by a perforated tape.

Concluding these introductory remarks, I would like to suggest the following criteria that might be helpful in deciding whether a function

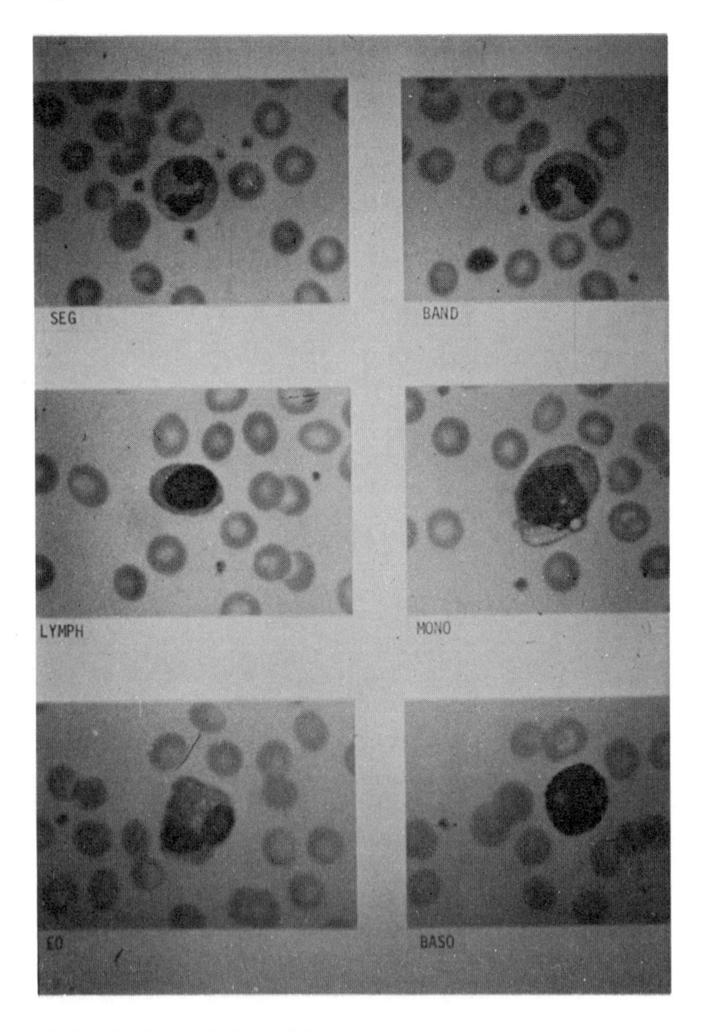

Figure 1 Morphology of Normal Leucocytes

of a clinical instrument or system should be computerised:

a. Computerising as a necessity. (A leucocyte classifier will be discussed as an example.)

and

b. Computerising as a means to add real value to a clinical instrument or system. (A bloodgas analyser will be discussed as an example.)

COMPUTERISED LEUCOCYTE CLASSIFICATION

Suppose that in clinical laboratories there was a need to classify automatically the normal six types of leucocytes and to print out their distribution in peripheral blood as requested for the differential blood count: the task of developing such an instrument could not be done without a computer. As can be seen from Figure 1 the types of leucocytes can be be identified by their differnt morphological structure and therefore, a computer can be used to classify them in a multidimensional decision space with features such as shape and size of nucleus and cytoplasm, colour, granularity etc. This is done by digitising each cell as shown in Figure 2. For Corning's LARC-system (Leucocyte Automatic Recognition Computer) a PDP8M processor from Digital Equipment Corporation performs the cell identification and some other functions. Before a brief description of this system is given, it is affirmed that the design philosophy

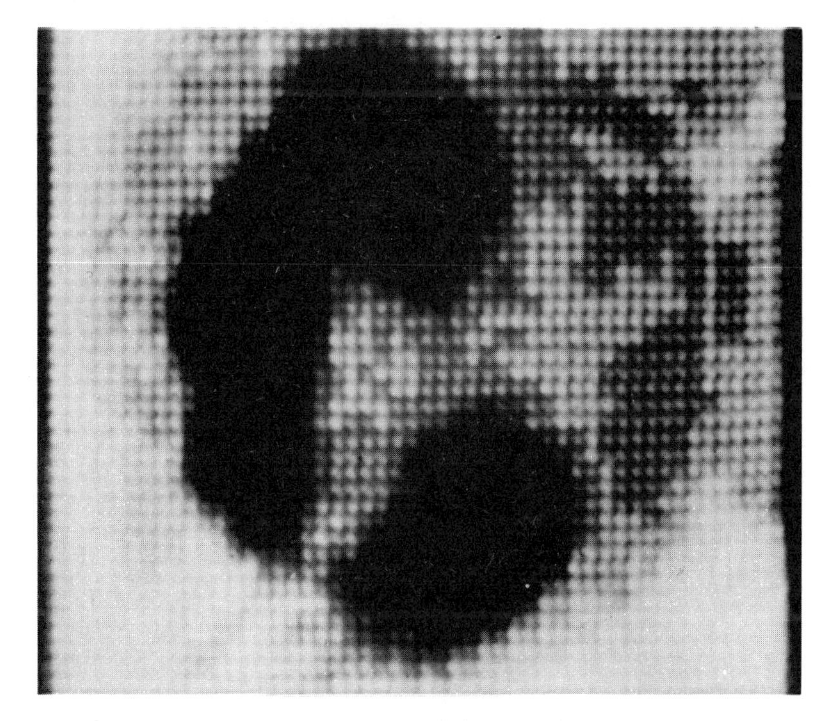

Figure 2 Digitised Image of a Blood Cell

of this clinical instrument falls under the definition 'a' described above.

With this system the automation of the differential count has been optimised and its performance constitutes an important step towards the automation of diagnostic haematology and cytology since it relieves the technician from time-consuming and fatiguing routine work and introduces a high measure of consistency into the evaluative process. The system consists of a computerised microscope that automatically classifies the following six white blood cells: segmented neutrophils, band neutrophils, lymphocytes, monocytes, eosinophils, and basophils. During operation, it also flags and counts all abnormal cells. The classification is performed by using cellular morphological features as used by a haematologist. If he wishes, the operator can manually view red blood cell morphology while the analyser performs the differential count and he can stop the analyser temporarily for making visual morphological judgments or to inspect abnormal cells in greater detail. Since during the automatic count, the x-y coordinates of every encountered abnormal blood cell are stored, the operator has the option of using the 'Review of Others' mode to view the abnormal cells after the differential count is concluded. In this mode only the abnormal cells are automatically acquired and focused, thus saving time which would otherwise be necessary for manual acquisition. The cells can be viewed either on the TV monitor or through the binoculars. More details of this can be found elsewhere (1-3).

This clinical instrument was designed in order to ease the tedious, fatiguing and time-consuming process in which normal leucocytes are viewed and classified manually; the design was purposely restricted to the identification of the normal types of leucocytes since the additional automatic recognition of abnormal cells as shown in Figure 3 would have resulted in an unacceptably expensive system. Inclusion of the morphological structures of abnormal blood cells and distinguishing their structures from non-avoidable artefacts was not considered part of the automation. Therefore, the system was not designed to replace clinical professionals but rather to give them more time to concentrate on the evaluation and diagnostic importance of abnormal cells. Since they are free from routine work for which a machine gives much better repeatability than a human will ever be able to achieve, they have more time for the difficult diagnostic activities.

BLOODGAS ANALYSIS

In this section an example is described for definition 'b', where added value was provided by computerising some functions of a known clinical instrument. After certain mathematical relationships (4) were developed and the significance of the acid-base physiology fully comprehended, the pH-bloodgas analysis became increasingly important for diagnostics and for therapeutic recommendations. It is the task of the acid-base relationship in our body to keep constant the hydrogen ion concentration of our body fluids. The pH values have to be kept constant in a very narrow range from which results the necessity and importance of an exact analysis. Important in judging the acid-base situation are the values of pH, PCO_2 and base excess. Of the many parameters, only the value of HCO_3^- is of significance since it describes the metabolic as well as the respiratory components whereas all other values do not permit their differentiation and are therefore of no diagnostic value.

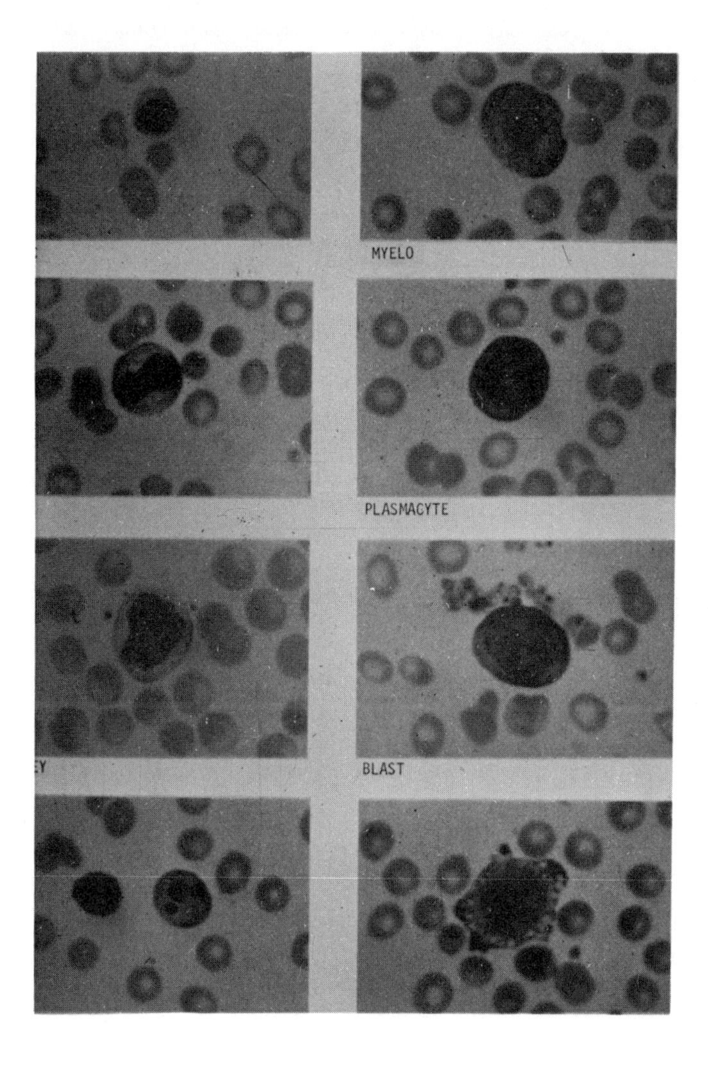

Figure 3 Morphology of Abnormal Leucocytes

In older bloodgas analysers the operator was required to calculate these parameters, make corrections, and to determine the sensor sensitivity and calibrate them from time to time. These functions are provided automatically in the new Corning bloodgas analyser by means of a dedicated microprocessor. The instrument uses a microprocessor to automate not only the measurement process but also to improve reproducability, precision and accuracy. The parameters pH, pCO_2 and pO_2 are automatically determined by calculating the endpoint of the sensor data after the blood sample is inserted into the instrument. From these three parameters four more are calculated, namely the bicarbonate concentration, the base and oxygen excess and the oxygen content. All data are displayed and the measured and calculated parameters are internally corrected if, for instance, the haemoglobin content or the temperature of the patient have varied. Automatic control of sensors, liquid levels, switches and the main parts of the electronic circuitry is included (5).

Table I Blood Gas Analysers and Microprocessing

Feature	*Without Microcomputers*	*With Microcomputers*
Operator interaction	Memorised by operator	Prompting with lighted displays
Calibration	Analog electronics, operator initiated	Digital electronics, automatically
Endpoint detection	Timed by operator judgement	Automatically from data analysis
Calculations	Analog electronics	Digital electronics using equations
Corrections (Hb, temp.)	Using charts and nomographs	Automatically on request
Diagnostics–Operator	None	Checks on membranes, liquid levels, and solenoids
Diagnostics–Service	None	Amplifiers, A/D converters, readouts, and indicators, computer, and computer interfaces

Table I compares in some details the features of a bloodgas analyser without and with a microprocessor, from which the differences in the overall performance can clearly be seen.

The sequence of operation is shown in Figure 4. For measuring the different parameters three electrodes (6), which are kept at constant temperature are used. The specification for the measured parameters is shown in Table II.

A block diagram displaying from left to right the functional work flow controlled by the microprocessor of the bloodgas analyser is shown in Figure 5.

With this application of modern microprocessor technology a device has been developed which is easy to operate, has self-calibration and self-diagnostic features and can be easily repaired because of error function displays. Since a relatively high degree of accuracy and precision is obtained (for instance, by automatic sensor drift correction) this clinical instrument can be considered a typical example where real value has been added with a microprocessor.

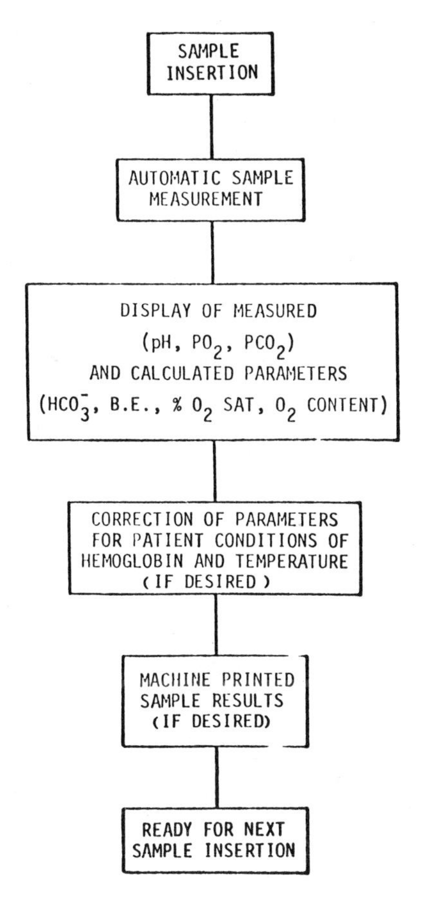

Figure 4 Sequential Diagram of the Bloodgas Analyser

Table 2 Parameter Specification

Parameter	*Range*	*Resolution*
pH	6.5-8.0	0.001 pH
PCO_2	5-250 mmHg	0.0 mmHg
PO_2	0-999 mmHg	0.1 mmol/l
HCO_3	0-54.0 mmol/l	0.1 mmol/l
Base excess (BE)	± 30.0 mmol/l	0.1 vol %
0_2 Content	1-64.0 vol %	0.1 %
0_2 Saturation	10-100 %	

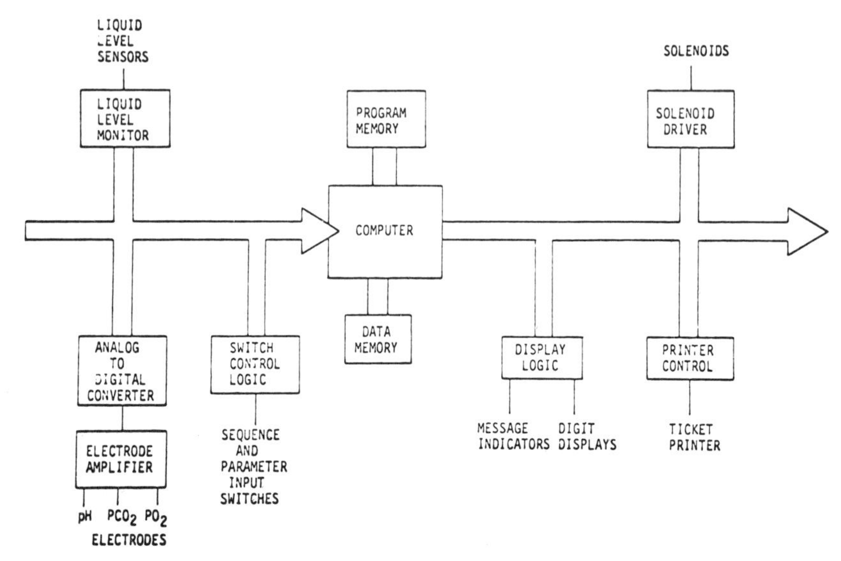

Figure 5 Microprocessor Controlled Functions

REFERENCES

1. Megla, GK (1973) *Acta Cytologica, 17,* 3
2. Cotter, DA (1973) *Am. J. Med. Techn., 39,* 383
3. Rogers, CH (1973) *Am. J. Med. Techn., 39,* 435
4. Siggard-Andersen, O (1966) *Ann. NY Ac. Sc., 133,* 41
5. Megla, GK (1976) *D. Med. Labor., 29,* 12
6. Severinghaus, JW et al (1968) *J. Appl. Physiol., 13,* 515

AUTOMATION AND 'POSITIVE IDENTIFICATION'

Marc Labarrere, Gerard Monnier, and Pierre Valdiguie
Departement d'Automatique CERT – ONERA,
Avenue E. Belin, 31055 Toulouse et
CHU de Rangueil, chemin du Vallon, 31054 Toulouse, France

INTRODUCTION

On account of the non-negligible risk of reversing samples in clinical laboratories so-called positive identification is a prime necessity and must materially improve the reliability of results.

Developing positive identification is closely tied to an overall automation of the use of analysers since positive identification is just one of the elements in the acquisition of data.

The microprocessor is particularly well suited to automation since it allows low cost computation. A system based on a microcomputer can be operated either as an autonomous unit or as part of a larger computer installation. Because it can be operated as a unit it is convenient to maintain. The functions it can provide could lead to the better use of non-specialist staff.

In our work, the addition of positive identification completes an existing automated system which has been developed for a continuous flow analyser.

The method of positive identification which we chose is based on the barcode. We are now investigating the effect of using barcoding on sample cups and the means of integrating the signals arising from an automatic reading of this code, with the other signals which are generated by a continuous flow analyser.

DATA PROCESSING

We will describe first the problems of the identification of a sample and then the processing of signals from the analyser.

Identification of a Sample

We use labels which are wrapped around analyser cups (Figure 1). The barcode is read in a vertical sense and because it is available around most of the cup, positioning is not critical so that cups can be quickly loaded. Another advantage of vertical reading is that no mechanism is necessary to rotate the tubes.

The 45 x 16 mm adhesive label carries the following information:

a. identification number in both plain language and coded form,

b. specific information stating the final operation station (Figure 2).

Some of the labels are kept aside to be used in special operations (e.g. calibration, dilution, control samples). So the dialogue between the operator and the computer is reduced to a choice of labels.

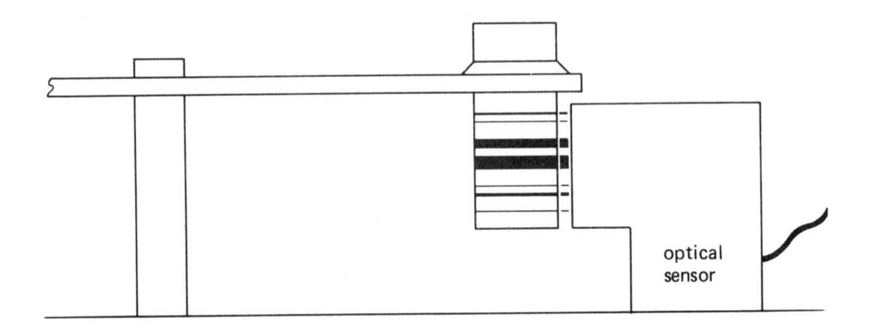

Figure 1. Analyser cup with barcode label which is read in a vertical sense by the optical sensor

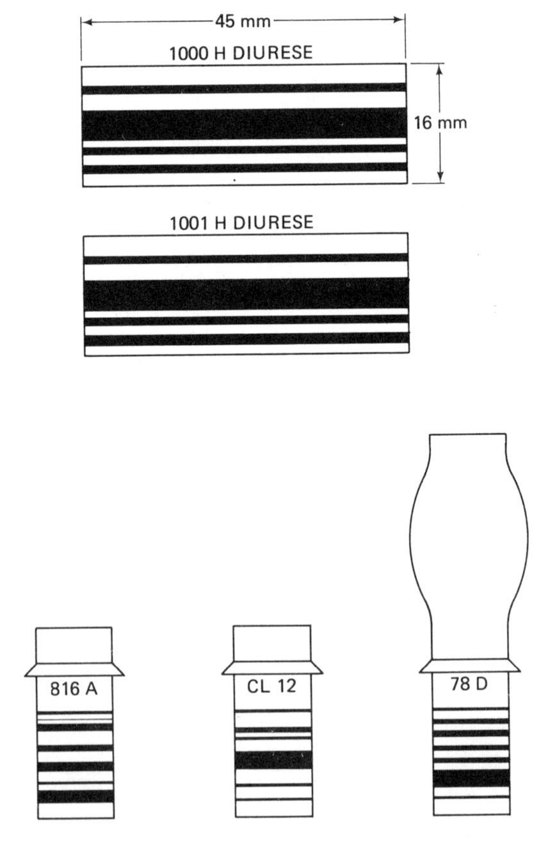

Figure 2. Barcode labels and analyser cups with labels

The sensor of the code is a photosensitive CCD (charge coupled device) element, made of a generator, and at least 256 photodiodes in a line and with a 3 mm photosensitive length. This apparatus gives an amplitude modulated analogue signal according to the state of the bits of the code to be detected. The analogue signal is converted by an ADC and this positive identification passes to the microprocessor (see Figure 3).

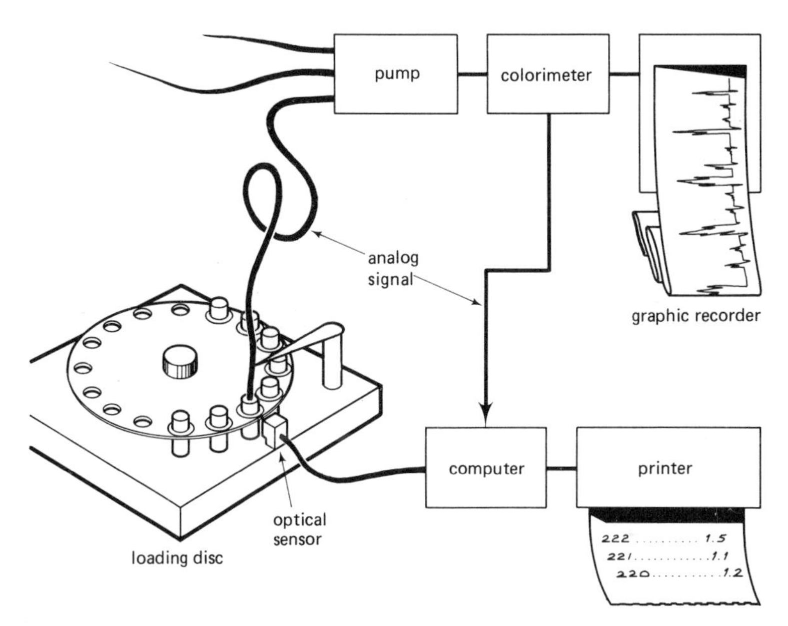

Figure 3. The computer system incorporating the optical sensor

Processing of the Signal from the Analyser

Because of the popularity of this kind of device in clinical laboratories we have chosen to use an autoanalyser. For this device we have designed an advanced method of control incorporating a microprocessor. The method can be applied to other kinds of signals issued from different units.

The variation of the elementary signal with time can be represented by an exponential function whose time constant depends on the characteristics of the analyser. The final value of the function is determined by the concentration of the sample. Thus the whole of the chain can be represented, as a model, by a first order system whose input is a sequence of steps corresponding to the different samples.

A Kalman type filter using an iterative least squares approach has been shown to be well adapted to the treatment of such well characterised signals. It is equivalent to looking for the nearest exponential time factor. It has two parts—one forecasting the next most likely value and the other

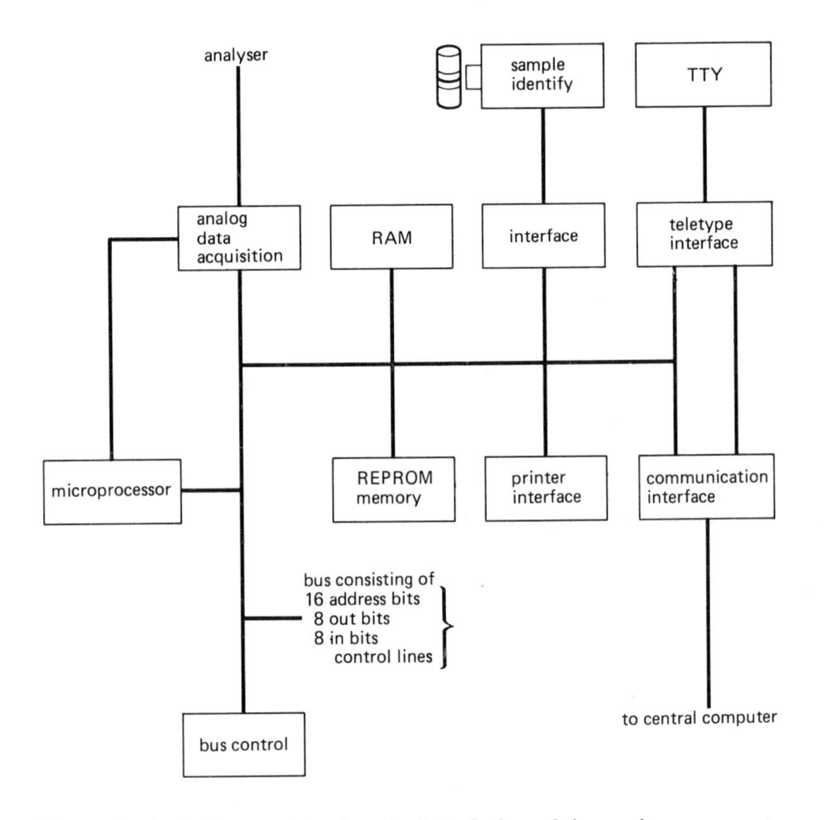

Figure 4. Block diagram showing the interfacing of the analyser computer components

an estimation derived by correcting the predicted value by using a function of its variation from the measured value.

The filter has two advantages. It allows the elimination of the interference impulses from the bubbles which are detected by large variations between the predicted and measured values, and the detection of the highest value (which is the value we are looking for).

Data Processing

Data processing includes the input of information, dialogue with the operator, printing of results at a local level and possibly communicating some information to a central computer. The processing has been highly simplified in order to reduce the time of calculation and to simplify its use by non-specialist staff.

Most of the information is sent with the help of the code on the label not only for prime samples but also for all the other operations such as diluting, checking and standardising.

IMPLEMENTING THE USE OF THE MICROPROCESSOR

The structure of the system is illustrated by the diagram (Figure 4) where the microprocessor is shown. It can exchange information through transmitting buses. The address bus gives out the address information in sixteen bits to the memories and to the peripherals.

The two data buses are unidirectional and are arranged in eight bit words. A control bus carries the different signals to allow the exchange of information.

The input set receives the analogue input from the analysers. It is connected to a RAM memory which can also be used as a working memory for the microprocessor. Thus input is made in an independent way through a memory and the microprocessor can also use the data kept in this memory.

In the REPROM memory is stored the MONITOR and the main program. This program carried out the functions of the introduction and the performance of any program held in the main memory, and also the display and alteration of the content of a memory at any address.

Peripherals for interaction with the operator include a small output printer and a five hexadecimal digit display unit with a 16 key entry for the input of common data required by the processor.

The control panel is not a peripheral used for operating. It is only used by either the programmer or the maintenance engineer to work directly to the buses by using machine code and for displaying the listings from the different lines.

CONCLUSION

In summary therefore there are two main points:

1. Positive identification is a part of the overall automation of analysers. Despite the present trend towards the marketing of digital devices for most of the analysers, it emerges that a microprocessor will be able to provide not only digital output but also positive identification or even further facilities at a low cost and with superior results.

2. The construction of new analysers using microprocessors will become increasingly modular. This will allow more decentralised arrangements which, in the end, will transform the analyser into a simpler form of digital detector.

ACKNOWLEDGEMENTS

The helpful criticism of the manuscript by Dr Martyn Roebuck is gratefully acknowledged.

COMPUTER NETWORKS AND MICROCOMPUTERS IN CLINICAL LABORATORIES

K Killian

Today many industrial process control applications make use of several computers to carry out special tasks rather than using a single large central computer. In clinical laboratories the trend to multicomputer systems is based on the present development in on-line interfacing of different analytical instruments, the organisational partitioning of central laboratories into a main laboratory, the emergency laboratory and the outpatient laboratory as well as the refinement of the result presentation.

The on-line connection of the analytical instruments was performed in the first computerised laboratories by a process control unit with various analog or digital input-output channels as shown in Figure 1.

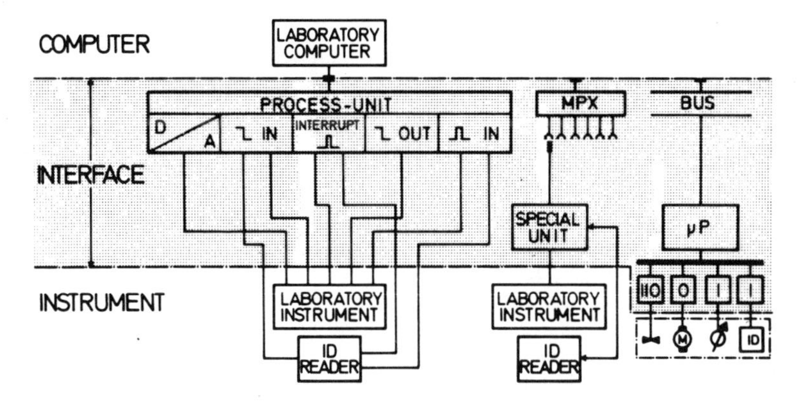

Figure 1. Examples of on-line connecting methods for analytical devices typifying earlier installations

In such a case signal management was by memory resident software which either worked by interrupts or by device polling. This procedure has several disadvantages. The time spent on hardware processing pushes the application software to a lower level and available memory space is used up for resident signal handling procedures. Besides, the selection of laboratory instruments was made without consideration of the software required to solve the interfacing problems. The extent to which hardware constraints affect software productivity was studied by William (1). Figure 2 shows that the software cost rises abruptly after reaching 85 per cent utilisation of hardware speed and memory capacity.

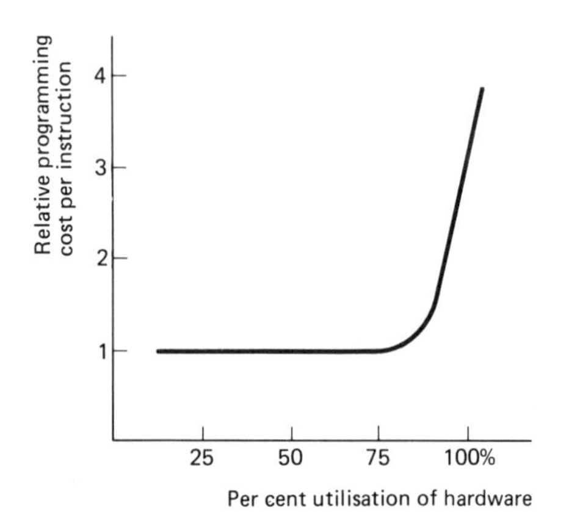

Figure 2. Effect of hardware utilisation on software productivity

For these reasons every effort has been made to concentrate the signal handling in special hardware interfacing units. The requirements for those units are:

acceptance of digital or analog test results,
signal and data control,
acceptance of status of the analytical device,
conjunction of patient identification and measurement,
addition of analytical method number and device number,
presentation of the transferred data,
input of user defined processing instructions,
output of control signals to the on-line laboratory instrument,
output via standardised hardware ports of uniform measurement result blocks consisting of measuring device identity, method identity, sample identifications number, device status, test result and parity control digit.

To reduce those problems the working group 'Laboratory data processing' of the German Society for Medical Documentation, Informatics and Statistics (GMDS) has in collaboration with numerous interface manufacturers laid down a universal specification for the hardware output ports (2). This hardware port is based in its serial version on one of the common line current ports (TTY). The parallel version incorporates a device data orientated handshake mode.

Classical interfacing was performed by hardware modules with fixed signal and data procedures. Now so called 'data preprocessing units' (SIEMENS, MVV 1009) use microcircuit switches or wired parameter connectors (Eppendorf) for this purpose.

The application of microprocessors allows two new approaches to solve the interfacing problems:

1. Hardware process control by use of PROMs for the communi-

cation basis program and for the external parameter supply. (DIA-log, DATA-CIN system)

2. Software process control by loading the control program into the RAM of the special instrument interfacing microprocessors (3).

An additional form of on-line connection is a direct incorporation of the microprocessor into the analytical device itself.

The development of microprocessor-controlled multichannel analysers with discrete test selection (KDA, American Monitor, PRISMA, LKB) or profile structure (SMAC, Technicon) is a direct approach to produce cheaper and more compact systems. Furthermore the purpose of biochemical instruments is not only to analyse the incoming specimen but also to present the results in a usable form to the physician.

The introduction of microcomputer systems into analytical devices has resulted in profound changes in their mechanical structure. Figure 3a shows the typical design of a 'classical' sampler (Technicon). The rectilinear motion and the rotation of the dispenser needle combined with manual

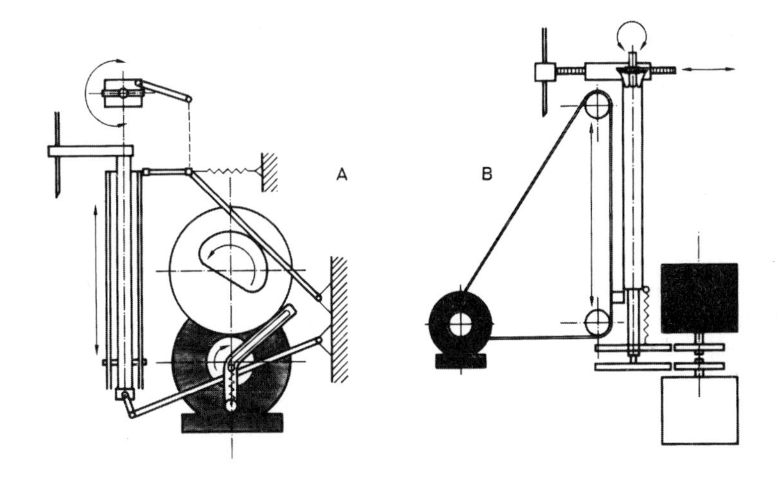

Figure 3. Mechanical design for a computerised sampler

vertical displacement is performed in this case by a motor and a semicircular disc. In contrast to this fixed lever drive Figure 3b shows a solution with three step motors for the X, Y, Z motion. In this latter version the timing and the displacement of the dispenser needle is controlled by a microprocessor. Furthermore, the mechanical parts required are reduced, the availability is improved, and the flexibility is enlarged.

This concept of splitting up the analytical device into its functional elements controlled by a microprocessor is exceptionally suitable for the design of automatic or semiautomatic distributors with direct sample identification. The wide variety of designs of the organisational structure in the laboratories requires a flexible distributing system. Figure 4 shows the main elements of an automatic sample distributing unit developed in cooperation with SIEMENS.

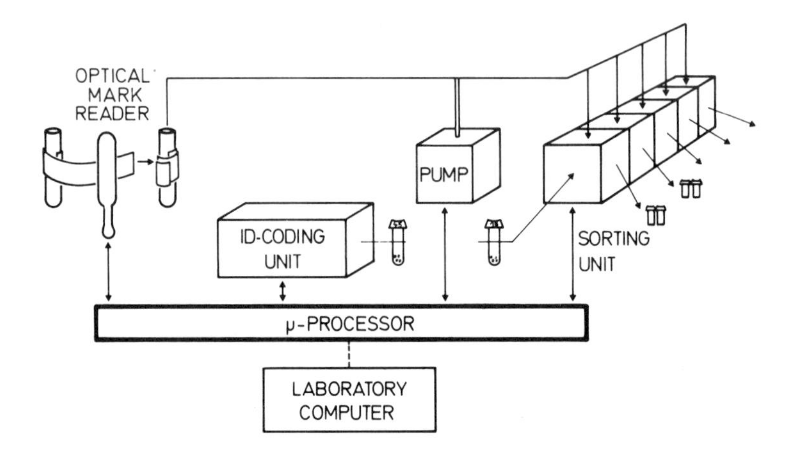

Figure 4. Automatic sample distributor with direct identification of the secondary vessels

The tests requested are transferred to the microcomputer by means of a special optical mark reader. The request form is drawn through this device by hand without separating the specimen. In this way the microprocessor receives all information to code, to fill up, and to distribute the secondary vessels. This design is prompted by the following considerations

1. The component parts of the sample distributor can be altered to meet the changing needs of the laboratory. As a first step neither the coding unit nor the pump are necessary, the microprocessor merely indicating where the manually filled vessels are to be placed. The coding unit and the pump can be added as separate items at a later time and the number of output channels can be altered at any stage of development. Modular software can keep pace with hardware changes by simple modification.

2. This dedicated computer system can also work in a stand alone mode and increases therefore the availability of the total laboratory system.

3. The structure of the request forms can be adapted directly for the user's need by means of position parameters.

As well as dedicated microcomputers of the type described above it is necessary to have a control computer for the cumulation and presentation of the test data. An outstanding problem is the best means of connecting the various computers in terms of both hardware and software. Five different elementary structures can be used as shown in Figure 5.

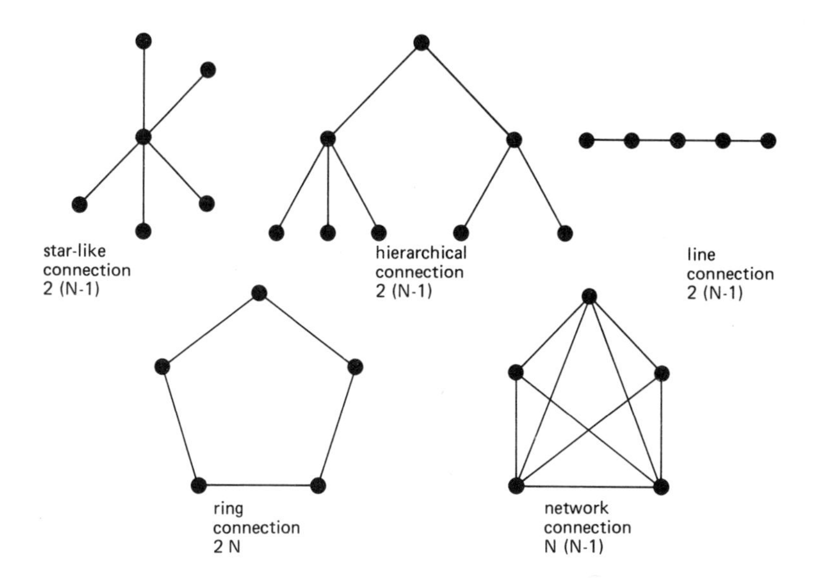

Figure 5. Multicomputer structures

The total cost of these different structures is a function of the number N of the participating computers. Figure 5 shows that star-like, hierarchical, line and ring structures have a cost which depends linearly on the number of processors. The network structure, which gives the highest degree of access between processors, has a quadratic cost relationship. The final implementation concept is therefore a combination of those elementary forms determined by the local environment.

A heuristic approach to building up a multicomputer system for a clinical laboratory is shown in Figure 6. Such a hierarchical network system has operated in our institute since 1976. Two special computers for the emergency laboratory and the out patient laboratory are included in the network.

The processing of all laboratory data is separated into two segments: data control and data presentation. The required program systems run on different computers (Siemens 330). The model works in stand-by mode for the data acquisition. The communication computer only stores all test results (drum) without any processing. The data control computer transforms these received measurements into final results in three steps by means of parameters as shown in Figure 7. In the first step a standardised measurement block is built up with the device specific parameters. In the second step the test orientated procedures are performed. Then the quality control including the preliminary results, the trend detection, and the correction of base line drift is realised. In case of errors the result block is transferred to the communication computer, otherwise it is

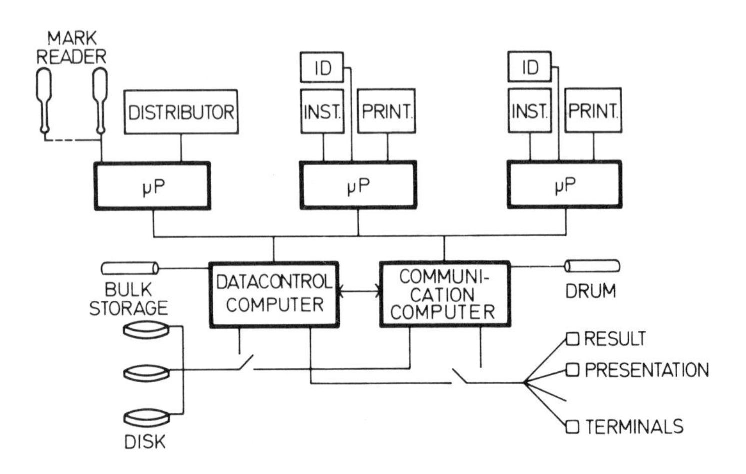

Figure 6. Multicomputer laboratory system in the University of Munchen Grosshadern

stored in the result file. At the end of a measurement series or initialised by external request, a fixed number of results are transferred to the communication computer for the purpose of recording in the patient file or output on printer or display.

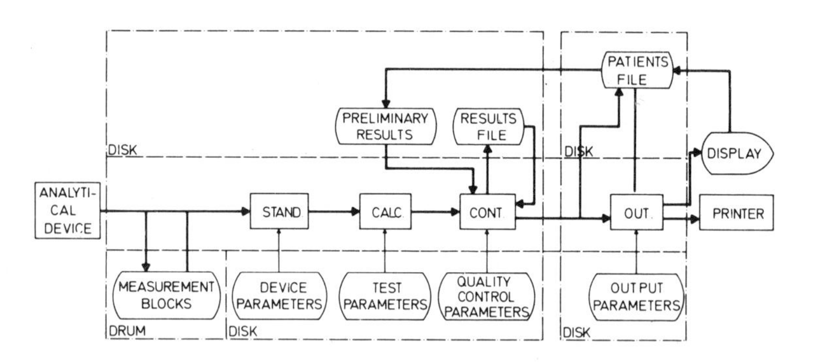

Figure 7. Data processing system in the multicomputer system Grosshadern

This approach to a multicomputer laboratory system is an open-ended concept. The hierarchical structure offers the possibility of dedicating the first level computers to rather standard functions for each analytical device. The central computer system realised in network structure is only used for high level data processing and communication problems with the special attraction of optimal availability. In our opinion the actual trend towards decentralisation of computer effort and the creation of autonomous working places in clinical laboratories is supported by this concept of a multicomputer system.

REFERENCES

1. William, A O and O'Donnell, C *Astronautics and Aeronautics,* July 1970.
2. Porth, A J, Killian, K and Pangritz, H, Hardware-Schnittstellen und Datenubertragungsverfahren fur den on-line Anschluss von Geraten im klinisch-chemischen Labor an Datenverarbeitungsanlagen, Arbeitsbericht 1 der Arbeitsgruppe Labordatenverarbeitung der GMDS, 1975
3. Blatt, E, Windrath, L and Winkler, E (1976) AESKULAP Anpassungs- und erweiterungsfahiges System fur Krankenhauser, Universitatskliniken und Laboratorien in arztlichen Praxer, Scientific Control Systems, SCS-Form 1128-1

Part 3

COST EFFECTIVENESS

AN EXAMINATION OF SOME PROBLEMS OF EVALUATING THE USEFULNESS OF COMPUTER SYSTEMS IN A HEALTH CARE ENVIRONMENT–AN OPERATIONS ANALYSTS VIEWPOINT

Dr Barry Barber

INTRODUCTION

In some exhibitions of modern art it is possible to see elaborate mechanical contrivances in which the component parts may be moved by an internal motor or some external agent creating interesting visual or audio effects. Although it would be a rash person that described such art as purposeless, these machines are in a strictly mechanical sense purposeless. The artistic purposes relate to the human observer and not to the mechanical effects of the machines. These devices contrast vividly with the sometimes similar looking machines in a science museum. The museum machines have been superseded by more efficient and effective machines but they are, nevertheless, clearly purposeful. The development and widespread introduction of water and steam operated machinery depended on an understanding of basic physical principles, the availability of an appropriate level of associated technology, economic and social pressures to achieve certain results and the inventive genius of individual scientists and engineers.

This historical aside serves to underline the basic stance of the operations analyst. He is inevitably concerned with purposes, systems for achieving them and ways of improving the efficiency or effectiveness of these systems. By their nature these systems are either man-made or there are human objectives in manipulating a natural system to achieve human purposes. The discipline of operational research, or operations analysis as it is more descriptively called, is not merely descriptive of relationships but is undertaken in order to make something in some sense 'better'. The operations analyst seeks an *analysis* of basic goals and attempts to model the most important parts of the system. This approach contrasts with the traditional activities of the systems analyst whose function tends towards a *synthesis* of a multitude of related activities by different individuals concerned with different aspects of the system. Undue simplicity in the work of an operations analyst can usually be corrected by more extensive data collection and more complex modelling. However, the same fault in a systems analyst may lead to a totally unworkable computer system, if important functions and uses of the system have been ignored. There is a tendency for computer systems to develop the complex character of multi-objective human systems which have grown by accretion, taking in lots of peripheral activities and subsystems, simply because it was possible to do so, and to help make it acceptable to as many people as possible. The ultimate failure of such a computer system is of course that no one would use it! This fact tends to obscure the basic systems objectives and outputs.

PROBLEM DIAGNOSIS AND PATIENT BENEFITS

Since the operations analyst cannot start his basic activity of mathematical model building until he has clarified the systems objectives, this diagnostic activity is necessarily one of his most important skills—much as diagnosis and decision making in medicine is an important precursor to patient treatment. As with the medical situation, there are similar dangers of treating the symptom of some organisational disease rather than the disease itself and, of course, of using some much loved mathematical manoeuvre as an all purpose panacea in inappropriate situations. The art of problem diagnosis, and at the present time it is still very much an art in the medical and health care field, involves a dialogue with individual members of the organisation somewhat as a clinician has a series of consultations with an individual patient and members of his family. As with a good general practitioner this is a continuing process punctuated by periods of acute or chronic activity as particular problems emerge and are dealt with. It would be wrong to pursue this medical analogy too far but there are some basic similarities and these can help to illuminate the situation of an operations analyst to an audience familiar with the practice of medicine.

Although it is undoubtedly an oversimplification to regard private industry as having only a single objective, the environment in which it functions depends on securing an adequate cash flow from which all other aspects follow. The analysis of patient benefits from health care procedures would provide a similar unifying basis within the Health Services but unfortunately very little work has been done in this area to date.

The problem of evaluating computers within the context of health care computing has little to do with computing *per se;* it arises as a result of the lack of evaluation of other activities within the health care services. The computer aspects are often complex and interrelated but at least in principle the detailed inputs and outputs to the computer systems can be adequately described. The basic difficulty lies in the assessment of the implications of these inputs and outputs on the functioning of the health care system or on the care of patients. These difficulties obviously are most acute in relation to multipurpose real-time information systems aimed at improving a number of basically administrative activities rather than systems concerned with a single, patient-orientated activity such as radiation treatment planning, intensive therapy unit monitoring or e.c.g. analysis.

HEALTH SERVICE PLANNING

The problems described above are multiplied when one considers the wider field of health service planning. Many of the attempts at computer systems evaluation have raised some of the wider issues and are leading to the realisation of the need for detailed technical work on the non-computing health care systems in order to clarify crucial planning issues. Indeed, the development of these moves towards health care evaluation may in the long run be of even greater significance for the health of the population—or indeed of individuals within the population—than the various implemented computer systems themselves.

Although planning has been undertaken within the Health Services, it has been related mainly to the implementation of a particular pattern of health care in some segment of medicine that was then thought in some sense desirable—usually issued as the report of some professional or multi-professional working party. Some reports have led to major changes and developments, others have been quietly ignored; some have required substantial resources for their adoption and then been unattainable with the resources made available and others relate simply to good professional practice.

Undoubtedly these periodic reviews of parts of the Health Service have provided foci for new development but unfortunately this is not often coupled with any serious evaluation of the proposals in terms of the benefits to patients, of the real costs of the proposals (as opposed to the out of pocket costs in a budgetary sense) or any cost benefit analysis to explore the relationship between likely, or proven, benefits and these costs.

In these cirumstances much of the planning in the Heath Service is carried out on the basis of input norms relating to the provision of services for populations, rather than on the basis of aggregated outputs from the system. Fanshell and Bush (1) and Rosser and Watts (2) have, among others, carried out some studies aimed at getting some output measures but this work is far from the stage of providing the basic data for planning purposes.

As with health care computing, but on a much larger scale, health care planning is beset with the problems of medical and health care objectives, efficiency and effectiveness. The reorganisation of the Health Service in England gives Regional and Area Health Authorities the responsibility for providing comprehensive health care for geographically defined populations. This enables them to make a start on the examination of their provision of services in relation to their populations, where the previous arrangements in which the teaching hospitals were administratively separate from the Regional Hospital Boards made such planning hopelessly complex if not impossible.

The attempts at the evaluation of computer projects have at once underlined the difficulties and the importance of carrying out adequate evaluation. Until some of the relationships between population health (i.e. morbidity and mortality) and the level of provision of various health care services have been defined, this planning is restricted to a geographical consideration for the provision of such services rather than an adjustment of the relationship between the services to provide some maximum patient or population benefit. However, it may be expected that the attempt to plan effectively will gradually uncover some of these relationships.

PROTOTYPE SYSTEMS

During the past 20 years computer systems have developed from obscure research and administrative tools into routine apparatus used widely for a large variety of purposes within the Health Services (3). In some cases these systems impinge directly on the central activities of patient care while in other cases they are concerned with the rather more peripheral administrative environment within which care is provided. Initially, many systems were in the nature of prototypes designed to explore the possibility, and the advantages of using computers in new application areas, to examine the acceptability to various groups of staff in the hospital and health care environment of the various types of system and available hardware, and to explore the sorts of hardware and software combinations which could be made to work in such unusual environments.

This process of exploratory system development has ranged from small scale minicomputers to substantial installations handling large quantities of patient information for record or analytic purposes, or communications systems to improve the information flow in a multidisciplinary environment. The small scale systems have tended to be rapidly and readily focused on specific problems and have led to systems which have now been very widely implemented. However, the larger systems have had a much larger impact on their parent organisation, but they have taken very much longer to design and build in a fashion acceptable to their organisation. Also, they have initially been focused on the relatively non-controversial preliminary administrative aspects of patient care rather than on the critical medical and nursing activities. The development of health care systems is illustrated in the systems hierarchy in Figure 1 (taken from Barber (4)). To date systems development has been proceeding from the peripheral to the central activities of medical care.

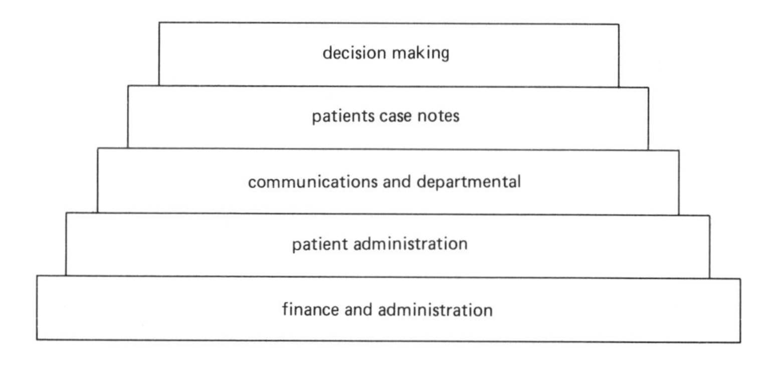

Figure 1. Development of Health Service computing systems

Many of the early attempts at handling medical records foundered because appropriate hardware and software facilities were lacking, because an appropriate strategy had not been devised for the integration of the new development within the organisation concerned (the management of change was inadequate and led to a transplant rejection), and because a number of the preliminary administrative systems had not already been implemented and which then deflected effort from the main records function. In the medical decision making and diagnosis, a number of relatively small scale studies are being pursued and should provide the basis for appropriate systems when the records systems have been fully developed. This process will, of course, be somewhat interactive because the records have to be related to development in medical decision making.

INFREQUENT SYSTEM USERS

Health care computing has involved developing systems for a most unusual computer user–the doctor, nurse, medical scientist or administrator who needs to use the systems for a small proportion of his time to gain access to its facilities but yet who cannot consider undertaking extensive training in computing science. Such training was accepted by many scientists in an earlier decade in order to pursue their research more effectively by gaining a detailed knowledge of computer hardware, machine codes and autocodes. However, these scientists were specialists who used, and expected to use, the computer systems extensively, at least for a substantial period of time. Most of the health care staff are concerned with system usage not system building and by now the research needs of such people are by comparison with earlier times well handled by existing statistical, survey and other packages and by ready access to machines using high-level languages.

Medical students, in common with many other students, now receive some basic introduction to computers as part of their training (e.g. Evans (5)). Nevertheless, considerable care has to be taken in the design of computer systems for infrequent users to ensure that they are easy to use in practice, are virtually self explanatory and require little training to enable staff to make use of their services effectively.

Once systems expand beyond the confines, and control of single departments, considerable effort has to be devoted to staff involvement and education. The elements of these requirements are now fairly well recognised and agreed. (A session at the medical computing congress MEDCOMP 77 held in Berlin in February 1977 had three speakers from different countries and disciplines independently emphasising the same basic points. Two of these papers are available in the proceedings (6).) Also there is a growing awareness of the problems of the management of change within the Health Services (7) which is of consequence for the whole range of health care planning activities and for the implementation of operational research in this field.

CLARIFICATION OF SYSTEMS OBJECTIVES

In order to evaluate computer systems, the objectives must be clearly identified and the system performance monitored, and costed, against these objectives. The evaluation of the system cannot be undertaken precisely if the objectives are hazy and ill defined, or if important objectives are not explicitly acknowledged. This is particularly important in relation to development programmes such as the experimental computer programme of the Department of Health and Social Security in England. This programme set out to produce a number of medium-sized, operational, real-time computer systems serving some hospitals, health centres and general practitioners, in order to explore their value to the Health Service. This main, and overriding objective, has been achieved and these systems have provided the basic information and experience required to design and build systems for wider implementation. However, the policy adopted by the Department on economic (or application) evaluation, indicated briefly by Sharpe (8), was misconceived and ignored the state of development of the subject. At that time this was very much at the prototype, 'will it work?' stage and the evaluation approach was not directed to exploring the major benefits that would, or indeed could, accrue from the programme. This error was compounded by attempting to subdivide applications to an excessive level of detail and by attempting to relate prototype system costs to observed benefits.

During the period of the programme the cost of computer hardware has dropped substantially and more appropriate software is available for system construction. The appropriate economic assessment should involve an estimation of the costs of short production runs on modern equipment using present day software to accommodate groups of applications as operational systems (not research or development systems) within a minimum hardware configuration. The benefits of the prototype systems should be reviewed in the light of these estimated costs and the likely achievement of benefits at a proposed site, in order to balance estimated costs and benefits for a potential purchasing decision.

The Department of Health and Social Security is currently reviewing its policy on evaluation and it is clear that many of the objections will be accepted and that the approach will be radically changed. An overall review of the programme has been outlined by Barber, Abbott and Cundy (9) as an approach to programme evaluation. This paper also indicates the role of project evaluation, technical evaluation and application evaluation, but the basic difficulties still relate to the questions of objectives and purposes.

EVALUATION OF MULTIPURPOSE COMMUNICATIONS SYSTEM

It would be wrong to generalise too much from the experience of evaluation of the London Hospital project (10) but many such systems are based on the patient administration approach and currently provide improvements in organisation, in the access to data and in its completeness, legibility and accuracy. Such systems frequently make *localised staff savings* in specific offices such as the admissions office, the laboratory offices and the medical records department and *distributed staff savings* throughout the organisation in initiating action or accessing information. The localised savings often involve large percentage staff savings in a very limited area as a result of the radical system design but these savings still only involve a small absolute number of staff. The distributed staff savings arise in two ways. First, the staff in a wide sector of the organisation can carry out certain functions faster. Second, by having direct access to information, the staff do not need to make enquiries of other staff, thus saving time overall in the organisation's (manual) information system. These distributed staff savings are difficult to measure with precision and involve small percentage savings for a large number of staff which can aggregate into a relatively large absolute number of staff. At the present stage such systems have yet to have a proven major effect on patient care and it is difficult in addition to evaluate with confidence the value of such distributed staff savings. Given this overall difficulty in measurement and the difficulties of allocating such savings to very small subdivisions of the operational computer system this approach can be prohibitively difficult if not totally unprofitable.

MANAGEMENT ACTION

Cumulatively distributed staff savings provided the opportunity for management action to translate these potential savings into real savings by appropriate reorganisation or by enabling the organisation to handle a greater work load with the same staff. Equally such benefits can be dissipated throughout the organisation and lost. The group working at El Camino Hospital evocatively term this approach to the process of obtaining practical benefits as their 'benefits realisation process' (11). Detailed studies have suggested that the financial savings can be well in excess of the computer system costs (12) and the most recent reports suggest very substantial savings by comparison with comparable hospitals without such comprehensive information systems (13). Norwood (14) has also drawn attention to the potential that such systems have in monitoring the quality of care provided by the hospital.

PROTOTYPE SYSTEM COSTS

One of the problems of the evaluation of the Department of Health and Social Security experimental computer programme in England has been that of setting prototype system costs against information intangibles such as legibility, completeness and accuracy and distributed staff savings, which have not yet been translated into real savings by management action, all of which have a small proven effect on patient care. The prototype costs are irrelevant apart from the initiation of a project. What is relevant to the evaluation is the cost of the production model systems, i.e. the cost at which such observed benefits might be purchased should they be required by some other institutions.

Much of the initial work was based on patient administration systems which provide an essential framework for medical care. Initially such systems were additional to the other health care systems but slowly these are being developed to handle more directly medical, nursing and scientific activities and thus they begin to impinge more and more directly on patient care and its high costs. Such systems will clearly develop to handle the patient's case notes and provide assistance for doctors and nurses with their information needs regarding patient care. There is every reason to think that the cost of production model systems will be much more closely in line with the overall system benefits in the short run and that the subsequent medical systems will be highly cost-effective.

EVALUATION OF SINGLE APPLICATION SYSTEMS

Turning now to substantially single purpose or departmental systems such as laboratory or radiation treatment planning systems, the situation is much easier. Firstly there is greater clarity in the specific requirements of the systems and they are generally central to some aspect of departmental activity. This opens the way to a cost equation with some basic measure of this activity leaving peripheral improvements in data access and legibility as additional benefits. Furthermore, since the system involves some central activity it is appropriate to assess the accuracy and performance of the system as a whole rather than attempting somewhat unconvincingly to trace the changes arising from the specifically computer parts of a major information system as compared with other changes in the manual medical, nursing and administrative aspects of the system over a period of years.

This can be illustrated with reference to some work carried out for the British Institute of Radiology and the Department of Health and Social Security (15). The approach attempts to separate the costs into those dependent on the volume of activity and those independent of it. These costs can at their simplest level arise directly from some rental or leasing arrangement or they can arise from the purchase of a system, its maintenance, running and staffing. In this more common latter case, some means will be needed of relating the initial purchase price to the recurrent running costs in subsequent years. This can be done in a variety of ways depending on the financial structure of the organisation involved. The simplest approach is to amortise the initial costs over an agreed period of years and add the resulting figure to the running costs. These calculations can be refined by using discounted cash flow techniques to bring the estimated stream of cash flows to a single present value and probability estimates may be included to give some idea of the likely variability of the final figure. Some of the running costs will be directly related to the degree to which the computer system is used for its main design function such as consumables and staff supervision time and the basic cost scrutiny involves the examination of all the costs in order to assess their relationship with this main variable.

In the case quoted of the radiation treatment planning system the main significant cost varying with the numbers of patients planned per year was the staff time taken to produce a radiation plan. To the first order of approximation the other costs were fixed. This led to the graph in Figure 2, which might be described as a one dimensional linear programme. In principle the operational characteristics of any radiation treatment planning system may be shown on the graph and a least cost system may be selected for any particular level of activity.

In this case, it has been assumed that all the relationships are linear but the basic concepts are not altered if much more complex relationships are observed. In the case of the computer system there was a basic setup cost whether or not it was used for its prime function and the operational costs per treatment plan were low. In the case of manual treatment planning there was no additional setup costs but the staff time per plan was high. For the batch systems their characteristics were not radically different from the manual systems; there were relatively low setup costs and the computing costs were low but the staff time was not greatly different.

From Figure 2 it is clear that it is uneconomic to install a computer system for very low activity but that as the volume of activity becomes larger the value of the real-time computer system becomes more and more evident. Specifically, one can conclude that, considering the systems as equal in other respects, the manual system is most economic when fewer than 500 patients need planning each year and the real-time computer system is appropriate when the load is greater. A similar examination of batch computer systems found them useful in the range 300-500 patients per year. The details of this calculation have been overtaken by rising prices and technical change but the principle will hold good and it is unlikely that the break-even figures have increased.

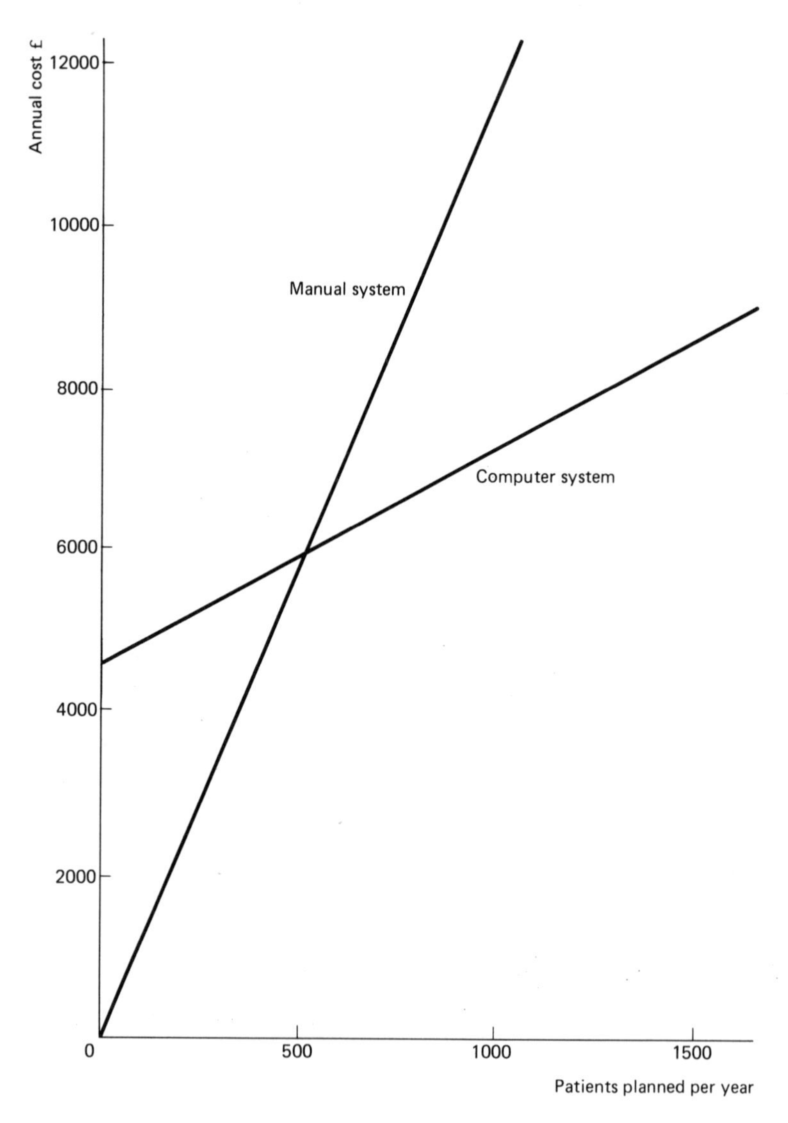

Figure 2. System selection in radiation treatment planning

This appreciation of the situation highlights the importance of a good assessment of the likely level of activity and in this case it involves esti-

for whom it is clinically or scientifically desirable to have individual treatment plans produced. This key variable sometimes gets overlooked in the discussion of computing requirements for medical systems because the discussion becomes centred around aspects of the computer hardware and software such as the relative merits of micro-, mini- and maxi-computers, of differing word sizes or of special facilities available with different operating systems or computer languages. These are important issues but they are mainly computer technicalities and require professional examination after the users requirements have been fully explored. The most important questions for the computer user and the evaluator are those relating to the volume of activity, the type of calculations that have to be made (e.g. does allowance have to be made for body inhomogeneity or for three dimensional plans?) and the speed and timing of the response.

This approach, also, enables one to take account of any other income, either real or virtual, which arises from the use of any spare time on the machine. This may function directly to reduce the initial setup cost of the system, thus yielding a parallel but lower system costline or it may be more complex as the additional system's use (and income) is reduced as the available machine capacity begins to be used up.

APPLICATION TO CLINICAL LABORATORY SYSTEMS

This illustration of system evaluation has been related to radiation treatment planning. However, these systems should readily yield to the same sort of analysis of objectives and the description of costs with respect to system activity. Some work would be required to determine the best descriptions of that activity, but numbers of tests or requests either direct or weighted are commonly used in Health Service statistics. Such basic information will be decidedly useful for functional budgeting as well.

I have seen this approach used in radiotherapy (16) but not elsewhere but it cannot be considered unique as it involves the fundamental logical structure of the problem—and, of course, it is the approach one uses in organising international congresses for budgetary control and fee setting!

REFERENCES

1. Ranshel, S and Bush, JW (1970) *Opns. Res. 18,* 1021
2. (a) Rosser, R and Watts, VC (1977) *Medcomp 77,* Online Conferences Ltd. Uxbridge, England, Page 559
   (b) Proceedings of Euro II held in Stockholm, November 1976
   (c) (1972) *Int. J. Epid. I,* 361
3. (a) J Anderson and JM Forsythe. (Ed) (1975) *Medinfo 74.* North Holland Publishing Co.
   (b) Collen, MF (1974) *Hospital Computer Systems.* John Wiley & Sons, New York
   (c) Laudet, M, Anderson, J and Begou, F (1976) (Ed) *Medical Data Processing.* Taylor & Francis, London
   (d) *Medcomp 77* (1977), Online Conferences Ltd. Uxbridge, England
   (e) *Medinfo 77* (1977), (Ed) Shires, DB and Wolf, H. North Holland Publishing Co.
4. Barber, B (1975) In *Computer Information Systems in Health Care,* Sperry-Univac, London

5. Evans, SJW (1977) *Medcomp 77,* Online Conferences Ltd. Uxbridge, England. Page 751
6. (a) Haessler, HA and Cooper, CG (1977) *Medcomp 77,* Online Conferences Ltd. Uxbridge, England. Page 625
(b) Scholes, M, Forster, KV and Gregg, T *Medcomp 77,* Online Conferences Ltd. Uxbridge, England. Page 639
7. (a) Fairey, MJ (1971) *Case Studies in Operational Research in the Health Services* (1977), Operational Research Society, Birmingham, England
(b) Revans, RW (1963) *Standards of Morale: Cause and Effect in Hospitals.* Nuffield Provincial Hospital Trust. Oxford University Press, London
(c) *Hospitals: Communication Choice and Change.* Hospital Internal Communication Project seen from within. Tavistock Publications, London 1972
8. Sharpe, J (1975) In *Medinfo 74,* (Ed) J Anderson and JM Forsythe. North Holland Publishing Co. Page 137
9. Barber, B, Abbott, W and Cunday, A (1977) In *Medinfo 77,* (Ed) DB Shires and H Wolf. North Holland Publishing Co. Page 913 (paper offered to Medinfo 77, Toronto, Canada)
10. (a) Fairey, MJ et al (1973) *The London Hospital Computer System: A Case Study in the implementation of a major real-time system.* Proceedings of a one day conference held in The London Hospital Medical College Hall 27.11.73 and 24.4.74, The London Hospital
(b) Abbott, W (1975 Planning and Implementation of The London Hospital Project In *Computer Information Systems in Health Care,* Sperry Univac London
(c) Barber, B (1975) In *Medinfo 74.* (Ed) J Anderson and JM Forsythe North Holland Publishing Co. Page 155
(d) Barber, B, Cohen, RD and Scholes, M (1976) *Med. Inform. 1,* 61
11. Norwood, DD (1975) In *Medinfo 74.* (Ed) J Anderson and JM Forsythe. North Holland Publishing Co. Page 149
12. Battele Columbus Laboratories *Evaluation of the Implementation of a Medical Information System in a General Community Hospital,* United States National Technical Information Service, Springfield, VA 22161 reference PB 248 340
13. (a) Private communication from Dr Richard Dubois (US National Center for Health Services Research) pending publication of detailed cost regressions of hospital costs
(b) Barrett, JP (1977) In *Medinfo 77,* (Ed) DB Shires and H Wolf North Holland Publishing Co. Page 973
14. Norwood, DD (1977) In *Medinfo 77* (Ed) DB Shires and H Wolf. North Holland Publishing Co. Page 23
15. (a) Barber, B (1973) *Computerised Dose Computation:* Report prepared for the British Institute of Radiology and the Department of Health and Social Security, The London Hospital, London
(b) *Medinfo 74* (1975) (Ed) J Anderson and JM Forsythe. North Holland Publishing Co. Page 801
16. Sternick, EF (Ed) (1974) *Computer Applications in Radiation Oncology,* Proceedings of 5th International Conference on use of Computers in Radiation Therapy, Page 167

SOME METHODS USED FOR EVALUATION OF LABORATORY COMPUTER SYSTEMS AT CHARING CROSS HOSPITAL

T J R Benson, BA, BSc, MSc
Computer Evaluation Unit
Charing Cross Hospital, London

INTRODUCTION

The Computer Evaluation Unit at the Charing Cross Hospital was established in January 1974 in order to evaluate the computer applications being developed by the major Department of Health and Social Security (DHSS)-funded experimental computer project at the hospital. The main computer applications which are either operational, or will shortly become operational at the Charing Cross Hospital are chemical pathology, haematology, microbiology, patients registration, patient master index, waiting lists, inpatient administration and statistics.

This paper outlines some of the methods used at Charing Cross for the evaluation of laboratory computer systems. It is not possible in the space to do more than point out some of the basic problems that have been met and which have required special effort to overcome. The particular areas covered include objectives of evaluation, choice of evaluation techniques, analysis of laboratory systems, use of attitude research techniques, assessment of the quality of service provided and measurement of the costs of laboratory activities.

OBJECTIVES

Evaluation is carried out for two main purposes. These are:

1. *Evaluation as Monitoring* The continuous monitoring of any venture involving risk (such as the introduction of a computer system) is invariably a paramount consideration of those responsible for the venture (i.e. management at laboratory, district, regional or national level).

2. *Evaluation for Decision Making* Evaluation may also be carried out to meet the known or predicted requirements of those who will take decisions about the future use of computers within the health service. These two purposes are different but may be complementary to each other. However, if the evaluator is not clear why he is doing the study then the evaluation may be no more than an end in itself. The basic evaluation procedure is shown in Figure 1. The first step is to identify for whom the evaluation is being done, and whether the report is to be used for monitoring progress or as an input to decision making elsewhere. At Charing Cross the evaluation work has been done primarily for the benefit of future decision makers, but we have attempted to meet the local monitoring needs as well.

The next step is to determine what information should be collected. This step is fraught with problems since it is difficult to identify what information is of greatest value *now,* let alone what will be needed for taking decisions several years into the future. Furthermore, the information needs of decision-makers are not static but require frequent reappraisal to take account of new political and economic factors, changes in medical practice and developments in technology.

Having decided what information is required, this must be obtained using the most appropriate techniques available. Finally, the evaluation findings need to be communicated to the audience.

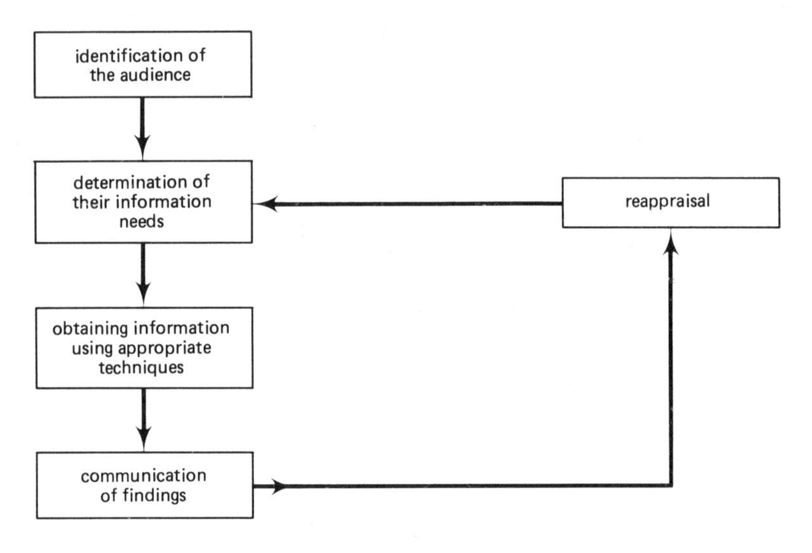

Figure 1. The evaluation procedure

CHOICE OF EVALUATION TECHNIQUES

Time spent evaluating some of the different measurement techniques available has invariably proved to have been well spent. At Charing Cross we have produced a list of criteria for choosing between different alternatives:

Suitability

is it relevant to the key hypotheses being investigated (the 'Improvement Objectives')?

does it enable comparisons to be made between measurements taken before and after implementation of the computer system, between applications in different health care units and between different applications?

does it measure performance by aggregating detailed measurements so that changes in overall performance can be attributed to their true causes (not necessarily the computer)?

Usability

is the technique readily available and accessible (for instance are special computer facilities required for analysis and if so are they available)?

is it well researched and soundly based on theoretical and practical experience?

is it within evaluation team resources of skills, time and money?

Support

is it well documented with instruction manuals, etc.?

are expert advice and/or training courses available, preferably on site?

are reports available of other peoples' experience and results using this technique?

Flexibility

is the technique limited to a particular task or to a particular hospital?

is it insensitive to the user? Would the same results be obtained when different researchers use the technique, or if it is used in a different environment? Will it cause the Hawthorne effect?

can we cover a number of aspects of the system (improvement objectives) with just one survey or study?

This list of criteria is formidable but it does show that an evaluator is unwise to attempt to design any technique from scratch (or to reinvent the wheel) without having first investigated all the alternatives. At Charing Cross we have found a number of useful techniques which score highly on the above criteria.

1. Standard workstudy and O and M methods such as described in Currie (1) for routine data collection,
2. 'Procedure for Determining Test Costs in Pathology Laboratories' developed by Cooper and Lybrand Associates for DHSS (2) for costing laboratory activities,
3. Questionnaires based on Man-Computer Interaction in Commercial Applications (MICA) Loughborough University (3) for determining staff attitudes to computer systems,
4. Statistical package for the Social Sciences (SPSS) (4) for carrying out statistical analyses of data collected.

ANALYSIS OF LABORATORY SYSTEMS

Before anything else can be done the evaluator must fully understand the system he is investigating. The work involved in doing this should not be underestimated; for example the evaluator must understand the similarities and differences between the separate laboratories, how they contribute towards medical care and how the different groups of people involved relate to each other.

Table 1 shows some of the key differences between the three main laboratories which directly affect the way that computers are used. These concern the nature of the tests done and data produced and the work/flow within the laboratory.

Table 1. Fundamental Differences between Laboratories

	Chemical pathology	*Microbiology*	*Haematology*
Workflow	parallel	sequential	parallel and sequential
Type of test results	numbers	words	numbers and words

In chemical pathology several tests may be carried out on different benches at the same time (i.e. the work flow is parallel) and test results are generally numeric. On the other hand, in microbiology one technician does most of the work on a particular specimen starting with appearance and microscopy and moving on to cultures and isolation of organisms; this is a sequential method of operation and the results are expressed in words rather than numbers. In haematology some of the work is reported purely numerically (e.g. FBC, ESR and platelets) and this work is done in parallel, but the film inspection is done subsequently (sequentially) and the results are often expressed in words. i.e. haematology has some of the characteristics of both chemical pathology and of microbiology.

The evaluator also needs to understand how the laboratories, and in particular computers in laboratories, may contribute to the care of patients. One simplified view of this is shown in Figure 2. The immediate effect of using the computer for processing chemical pathology, haematology, and microbiology results is to improve the reliability and availability of the laboratory test reports (i.e. provide better information). This may possibly lead to improved staff efficiency with a reduction of wasteful effort, clinical decisions more soundly based and/or more quickly taken, which in turn may lead to better management efficiency by the improved use of resources, improved scheduling of patients and/or improved patient care. If this is achieved then medical care can become more efficient.

It is also valuable to assess how the different groups of people involved in the system relate to each other. Figure 3 shows some of these groups and the direction of some important interactions between them. The arrows indicate the more important direction of influence. (N.B. All influences act in both directions but not equally.) The dotted line across the figure separates ward and clinic areas from the laboratory; it also separates the area where the laboratory management has greatest control. The number of arrows in or out of a group is a crude measure of the pressure the group is under. Perhaps it is the conflicting pressures on laboratory clerical staff (five arrows in; one out) and porters (three arrows in) that causes them to be the subject of most complaints. It may also provide the opportunity for the service to be improved by using a computer.

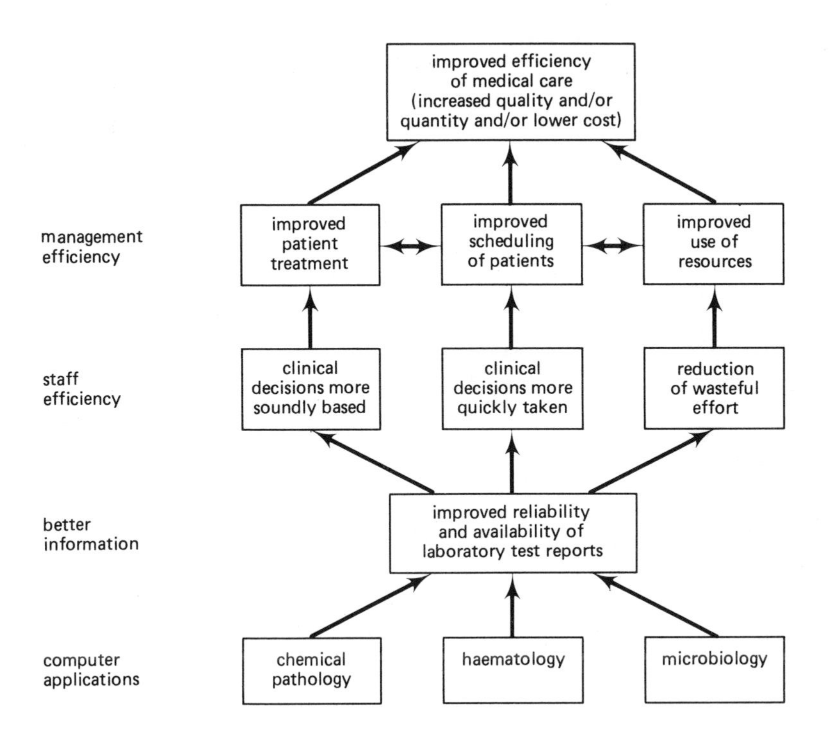

Figure 2. How using computers may affect the efficiency of medical care

USE OF ATTITUDE RESEARCH TECHNIQUES

At Charing Cross we have always taken the view that one important aspect of evaluation is to ask the people involved–the users–what they think about the system. Our initial work on the attitudes of laboratory staff was based on the MICA project work of Eason et al (3).

Respondents were asked to classify various aspects of the computer systems using five point scales typically ranging from 'very satisfactory' to 'very unsatisfactory'. The responses may be aggregated into four main factors: task fit, accuracy of information, ease of use, and user support, as shown in Table 2.

Unfortunately we fell into a common trap and did not use *exactly* the same questionnaire in each laboratory nor did we ask the questions of precisely the same sample of staff. This makes invalid any detailed comparison between laboratories.

These results are presented here simply to show the sort of results that might be achieved using a standard questionnaire and controlled

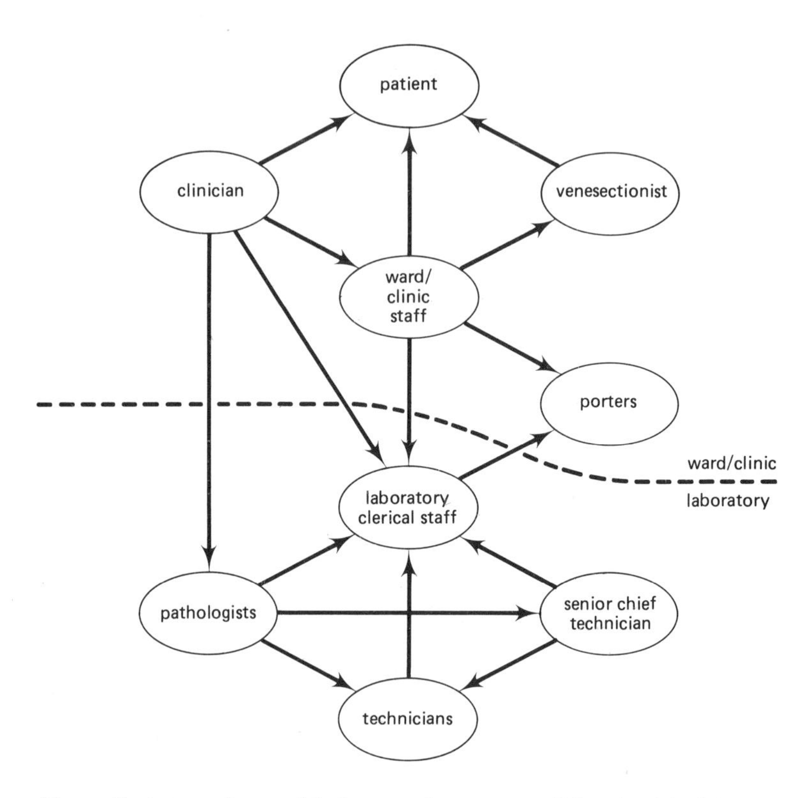

Figure 3. Interaction and influences between staff involved in laboratory tests

Table 2. Attitudes of Laboratory Staff Towards the Computer System

	Scores *a*		
	Chemical Pathology	*Haematology*	*Microbiology*
Task Fit	80.0	91.4	75.3
Accuracy of Information	84.2	90.7	80.0
Ease of Use	64.1	69.8	69.2
Support	70.0	60.0	72.0
Final Choice b	100.0	95.6	85.8

a Maximum score is 100, minimum is 20.

b The final choice question asked respondents whether they preferred to work with or without the computer or were indifferent.

samples. Comparative data such as this is of particular value in monitoring and controlling the introduction of computer systems. Many of the aspects of the Charing Cross system which received relatively low 'scores' in 1975 have since been the subject of further development by the computer project.

QUALITY OF SERVICE

The main objective of the computer systems is to improve the efficiency of the laboratory service. Unfortunately there are no simple unambiguous measures of the efficiency of a laboratory service, but it is a function of *the quality of service provided to clinicians and the cost of providing the service.*

Quality of service is particularly difficult to measure. At Charing Cross we have interviewed clinicians about their opinions of the service they receive from each of the laboratories. The interviews were conducted personally and took the form of a semistructured questionnaire allowing clinicians to express their views in their own words. The interviews were taped for subsequent study and analysis. Many of the answers were couched in comparative terms relative to the respondents' experience elsewhere or before computers were introduced. The interviews identified both the strengths and the weaknesses of the laboratory services and have provided feedback to the management of the laboratory which has been considered very valuable. However, the information generated by this sort of interview is purely qualitative and needs to be supplemented by other measures such as turnround time, etc.

Even an apparently straightforward measure such as *turnround time,* defined as the total time spent in the laboratory from reception of request and sample to despatch of the report, may be expressed in many different and ambiguous ways. For example if turnround time is expressed in hours then what conventions should be used for overnight or weekend delays? Turnround time is often easier to measure than *transit time* (the total time between request being made and the report returned to the clinician) but transit time is a better measure of the quality of service.

Some evaluators have attempted to circumvent these problems by expressing turnround time in terms of the percentage of requests having same day delivery of reports. However, this can also be ambiguous. For instance in Charing Cross chemical pathology laboratory the following statements (Table 3) were all true in September 1975 for tests requiring less than three hours bench-time.

Table 3

	Per cent of reports issued on same day
Requests received before 11 a.m. 'deadline'	100
Total requests received	66
Requests received after 11 a.m. 'deadline'	0

Depending on which figure is used one can give a good, middle or bad impression of the performance of the laboratory.

COSTS OF LABORATORY ACTIVITIES

In the haematology and chemical pathology laboratories we have measured the costs of key laboratory activities using the "Procedure for Determining Test Costs in Pathology Laboratories" issued by DHSS, 1975. The elements of cost which make up each activity are manpower, disposable materials, equipment and computer costs and overheads. Not all of these will apply to any one activity. Separate cost components are aggregated to activity costs and similarly activity costs are aggregated to request costs (Figure 4).

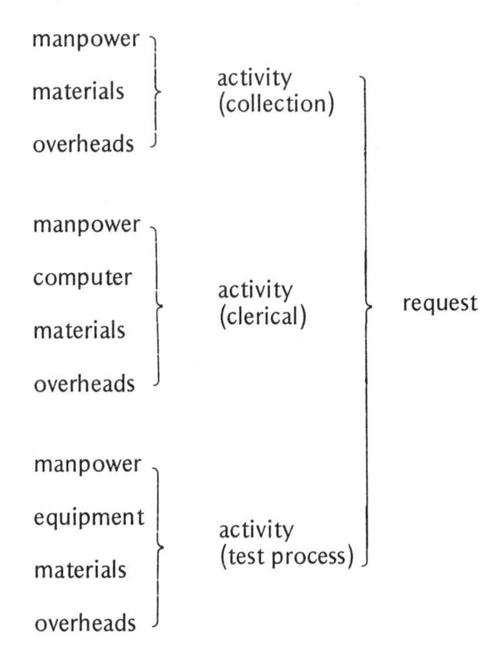

Figure 4. Structure of Request Cost

The procedure as used at Charing Cross Hospital generates a great deal of valuable information for each activity. This is shown as follows:

Number of items per week
Manpower – time per week,
time per item,
cost per item
Equipment – time per item,
cost per item,
maintenance costs,
depreciation

Disposable materials used
Computer–proportion of total use,
cost per item
Overheads–area occupied by activity,
cost per item
Activity cost per item

We have now used this procedure three times; for haematology in May 1974 and in March 1977 and for chemical pathology in November 1974. The exercise requires about three man months work by the evaluation officer and the full cooperation of the laboratory management. However, it is comprehensive and rigorous, and allows comparisons to be made both between laboratories and in the same laboratory at different times.

Table 4 lists the costs of some haematology activities at March 1977 prices, together with the percentage increase in cost over May 1974. Activities directly affected by the computer are shown separately from those unaffected by the computer.

Table 4. Unit Costs-Haematology Laboratory

Test Activity	*March 1977 (pence)*	*Per cent increase over May 1974*
1. *Activities directly affected by computer:*		
clerical work (manpower only)	17.7	45
full blood count	15.7	32
platelets count	22.1	-16
2. *Activities unaffected by computer:*		
film preparation	18.6	82
film viewing and differential	44.5	82
ESR	9.1	184
specimen collection and porterage	35.7	75

It must be emphasised that the differences between May 1974 and March 1977 are not between a manual system and a computer system but between two computerised systems. The main enhancements which have taken place are:

Clerical work Provision of computer printed day-book and results sheets, reducing the effort at specimen reception and answering queries. Entry of full blood count results is now not handled by clerical staff.

Full Blood Count The Coulter 'S' is now 'on-line' to the computer. In 1974 results were produced on punched paper tape.

Platelets Count An automated Technicon platelet counter has replaced the Coulter 'F' counter. This produces results on paper tape for entry to the computer.

ESR The large percentage increase in costs of ESR is explained by a very large increase in the cost of tubes and by differences in the time spent. Each test takes one hour elapsed time but less than a minute of technician time. This varies according to the workload and was higher in 1977 than in 1974.

FINANCIAL APPRAISAL

The most important result of any socio-economic evaluation of the sort described above is information concerning the financial viability of the computer systems being considered. A financial appraisal is essential to answer the basic question: 'Will the finance officer approve it?' This information is not easy to provide; for example the historical costs of systems in the DHSS research and development programme are misleading for decisions on current proposals for laboratory computers. Financial benefits identified in individual laboratories have depended greatly on local circumstances such as the scope of computer system, size of hospital and previous systems.

However, last year the DHSS conducted an 'Interim evaluation of the NHS Experimental Computer Programme' (5). This involved collecting evidence from some 23 laboratories at 12 sites and from this evidence it is possible to estimate the time saved by computer assisted systems, which may be valued at standard costs and set against the cost of providing the system.

First the scope of the computer system must be specified. We will assume a laboratory-based computer system covering chemical pathology, haematology and microbiology handling 5,000 requests per week. The facilities provided are to be similar to those of the Phoenix system developed by Professor IDP Wootton at the Hammersmith Hospital (6). However, it is stressed that the figures used are illustrative only and do not refer to any existing project.

Investment outlay required is £60,000 based on December 1976 costs for a computer configuration of 64k word memory and 20 megabyte disc storage with three VDUs and the printer in each laboratory (chemical pathology, haematology and microbiology). The annual operating costs include maintenance at say 8.3 per cent of first costs (£5,000), consumables, stationery, discs, etc., (£3,000) and staff–two operator/programmers and three data preparation clerks costing £17,000 a year.

Table 5. Potential time savings in laboratories

	Seconds saved per request/specimen	*Hours saved per 1000 requests/specimens*
Clerks		
Laboratory reception	35	9.7
Retrieval/creation of laboratory records	90	25
Production of reports and filing	72	20
	197	52.7
Technicians		
Worksheet production	25	6.9
Result calculation	60	16.7
Result transcription	72	20
	157	43.6

Potential time savings are shown in Table 5. These are marginal savings from using an efficient computer system compared with an equivalent manual system. It should be noted that much of the evaluation data used to derive these figures were obtained from laboratories in large teaching hospitals and may not be representative of smaller laboratories. The staff savings must be offset against the costs of computer operations and data preparation.

These savings are not achievable immediately on implementing a computer system; it is realistic to assume that one third of the savings are achieved in the first year, two thirds in the second year and all in third and subsequent years. Using DHSS standard costs for clerical staff and technicians (£3,073 and £4,881 p.a. respectively) and all the assumptions made above (e.g. 5,000 requests a week) a seven year cash flow may be produced (Table 6).

Table 6. Cash Flow for a Laboratory Computer System

Year	*Investment outlay (thousand pounds out)*	*Operating costs (thousand pounds out)*	*Staff savings (thousand pounds out)*
0	(60)	–	–
1		(25)	18
2		(25)	36
3		(25)	54
4		(25)	54
5		(25)	54
6		(25)	54
7		(25)	54

This may be shown more dramatically in graphical form (Figure 5). A discounted cash flow calculation may be carried out on these figures which gives a DCF rate of return over 20 per cent. This means that on financial grounds alone (ignoring all consideration of improved services) the laboratory computer system is a 'profitable' investment provided that money is available at interest rates of less than 20 per cent p.a.

The calculation is, however, sensitive to increased costs (raising the cost line) or to delays in achieving the benefits predicted (moving the savings line to the right). One consequence of this is that the development of one's own computer system requiring both extra development costs and subsequent delays can not be recommended except in special circumstances.

CONCLUSIONS

This paper has only skimmed some of the fundamental problems involved in evaluation of laboratory computer systems. The difficulties must not be underestimated. What is needed is much greater coordination of evaluation effort than we have seen in the past. It is essential that we should be able to compare directly the performance of different laboratories using measures which are acceptable to the management of the laboratories concerned.

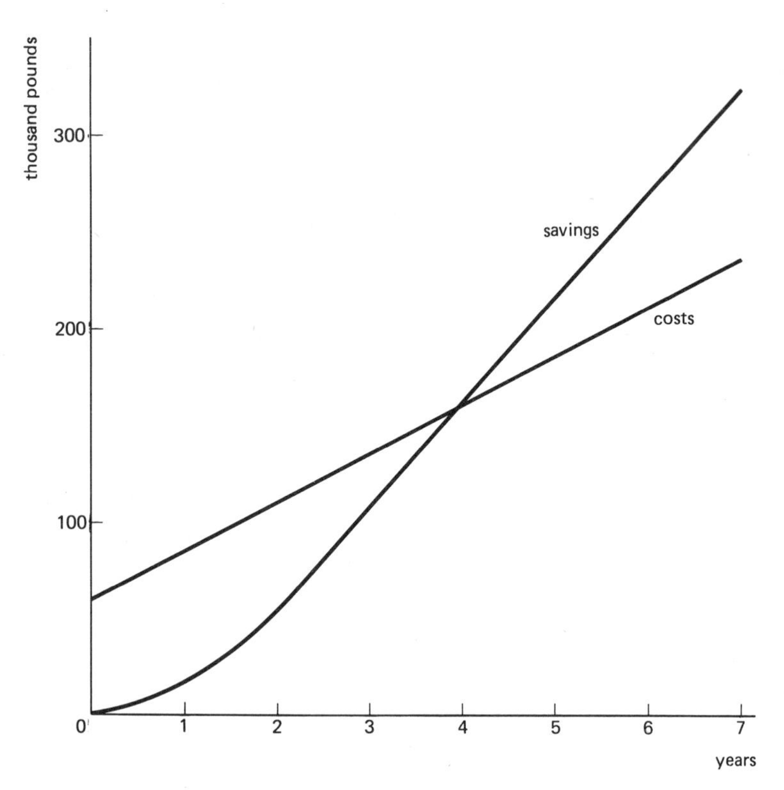

Figure 5. Cumulative cash flow

The confidence and cooperation of all concerned is only obtained by using an approach that is quite clearly thorough, impartial and above all objective.

ACKNOWLEDGEMENTS

The author acknowledges the help and support provided by DHSS, members of staff and management of the Department of Medical Computing Science, of the Hospital and District and of the Charing Cross Hospital Medical School and of the other members of the Computer Evaluation Unit.

REFERENCES

1. Currie RM (1973) *Work Study* Pitman, London
2. DHSS (1976) *Procedure for Determining Test Costs in Pathology Laboratories* London
3. Eason KD, Damodaran L and Stewart TFM (1974) *A Survey of Man-Computer Interaction in Commercial Applications* Department of Human Sciences, Loughborough University of Technology

4. Nie NH et al (1975) *SPSS* 2nd edition, McGraw Hill
5. DHSS (1977) *Interim Report on Evaluation of the National Health Service Experimental Computer Programme* Section 9 Laboratory Systems, London, Page 127
6. Abson J, Prall A and Wootton IDP (in press, 1977) *Annals of Clinical Biochemistry*

EVALUATION OF A HAEMATOLOGY COMPUTER SYSTEM

Donald Grant Chalmers, MA, MB, FRCPath
Department of Haematology,
Addenbrooke's Hospital, Cambridge

INTRODUCTION

Although my laboratory was first evaluated in April 1972 and again in July 1975, and, I hope for the final time, in July 1977, I cannot claim to be an expert on this subject. However, I do have as much experience from the recipients point of view as one is likely to find at present. In order to clarify my brief, I thought there might be advantage in attempting to understand the precise meaning of the term to evaluate. In order not to be too pedantic I chose the Shorter Oxford Dictionary: 'To evaluate: To reckon up, ascertain the amount of, to express in terms of the known.' I found that none of these definitions of the verb 'to evaluate' appeared to cover the work which we are considering today. I was very much attracted to the third definition, i.e. to express in terms of the known. Surely all concerned with computer systems must applaud this approach.

There is considerable difference in attitude between those laboratory scientists who view the development of laboratory computer systems with interest and those who are concerned to introduce a computer system into their department. In the first instance, much enjoyment can be gained by visiting one's colleagues, examining their systems and admiring or despising them at a relatively superficial level. The second group have a different task. In our present national impoverished and cost conscious state the concept of spending additional money in order to maintain or enhance an information system is not likely to appeal to our multi-tiered administrative masters. It is therefore necessary to obtain, wherever possible, hard facts and figures which support and justify the claims made on behalf of particular systems.

The difficulty lies in determining how the benefits may be quantified and assessed. In the early days, after the establishment of evaluation as a separate activity within the research and development programme sponsored by the Department of Health, the concept of the improvement objective was devised. This was a description of the improvements which were expected to be achieved by the introduction of a computer system

In terms of my own haematology system, five major areas were defined. These are as follows:

1. Effect on information
2. Effect on staff
3. Effect on patients
4. Effect on management
5. Effect on community

Each area was then broken down into a number of categories. An example of this breakdown is shown where six categories of effect on management are specified:

4. Effect on management
   (a) Clearer more accurate and informative cumulative reports
   (b) More precise results from Coulter S
   (c) Earlier warning of deviation of Coulter S
   (d) 'Better' results due to long term monitoring
   (e) Faster and more accurate reporting
   (f) More comprehensive and accurate statistical reports on long term workload.

Although this may seem a logical approach to the problem, it does carry with it certain dangers. It could be that concentration on the defined objectives which inevitably are selected before the exercise begins may divert attention from other benefits which exist but are not readily apparent. I would suggest, therefore, that the evaluation exercise cannot be left in the hands of professional evaluators but that heads of departments and laboratory managers must be involved and be satisfied that any improvement, however small, is noted and quantified.

There was general agreement in the early days of evaluation that improvements could only be assessed adequately if it was possible to compare data obtained before computerisation with that obtained afterwards; a number of exercises were set up where so called baselines or precomputer evaluation activities were carried out. It was because of this approach that my own department has been evaluated on two separate occasions. The first was shortly after the initiation of the development of the computer system(1). This was carried out by the computer development team assisted by officers of the Department of Health. Some eighteen months later, when the system was still under active development the department moved from accommodation in the old hospital into a much larger brand new laboratory complex in the new Addenbrooke's Hospital. Associated with this was an increase in staff and a whole range of new equipment including a Coulter Counter Model S which replaced the existing Technicon SMA 4. The entire pattern of work was changed. As a result the first evaluation did not present a reasonable baseline and it was decided to repeat it. This time the exercise was organised and conducted by the evaluation team based on the departmentally sponsored hospital system. This second baseline study was completed in July 1975(2).

Extensive discussions were held between the evaluation team leader and members of the laboratory staff to define not only the appropriate measurements to be made but the way in which they could best be recorded. Seven major areas for study were specified. These are as follows:

1. Laboratory staff
   - Size and grading
   - Activity sampling
   - Attitude survey
2. Workload
   - Size
   - Variation
   - Throughput
   - Turn-around time
3. Laboratory errors
4. Gross errors
5. Quality control
6. User attitudes
7. Cost

The first of these is laboratory staff. Obviously the total number of staff and the number in each grade must be recorded. As not all the laboratory, e.g. the blood transfusion section, is involved in the computer project, only those staff involved have been included in the survey and during the evaluation exercise changes of staff between sections did not occur. The working of the staff was recorded in a number of ways. Firstly, an activity survey was carried out. This involved an observer walking around the laboratory at five minute intervals and recording the activity of each member of the staff under one of a large number of headings. In this way an assessment could be made of the relative amount of time spent on instrument control, transcription of results, calculations, maintaining records, teaching, management activities, etc. This survey also applied to the clerical staff concerned with the main haematology laboratory. The use of the telephone, a major activity in the laboratory, was estimated by a self-recording procedure. Time and duration of all incoming and outgoing telephone calls was noted and associated with this was a code to define the type of message being transmitted. It is well known that the speed and quality of performance of staff of all grades is coloured by their attitude to the task in hand. This was assessed by the use of a semistructured attitude survey which was carried out, at a time close to, but not during, the evaluation measurement period.

A large amount of data was collected on workload. The difficulty of deciding the most useful information to collect is highlighted when consideration is given to turn-around time. What does turn-around time mean; to the clinician it is from the moment at which he collects the sample or indeed when he creates the request to the time at which the result is available to him. The haematology department may well have very little influence, if any, over the time which elapses between the generation of the request and the receipt of the specimen in the laboratory. Although reports may be produced for despatch, the despatching procedure is usually controlled by the hospital administration and again the laboratory has no control over the time taken for the result to reach the ward. Because resources were limited, recording of turn-around time was restricted to the moment at which the specimen appeared in the laboratory to the moment at which the report left the laboratory.

Laboratory errors are those mistakes in transcription, in failure to perform tests or in the production of a wrong result, which may be detected in the laboratory and corrected before the result is transmitted. Any incorrect result transmitted outside the laboratory is regarded as a gross error.

The creation of a quality control system in haematology is a very complicated issue. Because many tests used by haematologists must be carried out on fresh material, there is difficulty in providing permanent standards. For this reason haematology lags behind biochemistry in the breadth of quality control procedures used. Quality control is usually taken to mean the control of the blood counting equipment–in the case of this project, this is the Coulter Counter S. There is no regular quality control of platelet counts, erythrocyte sedimentation rates, reticulocyte counts, differential cell counts but in the cumulative system implicit in this project some quality control is built in since previous results on the patient are available for scrutiny at the time that the current result is approved.

Perhaps mistakenly, it was decided in the baseline study not to embark on major testing of user attitudes. It was realised that this could take a considerable length of time and the manpower resources were not readily available. It was also my opinion that I was aware of the range of complaints that were likely and that I could describe the nature of the complaints to be expected. It is also questionable whether a baseline and an after evaluation of user attitudes could be compared. Many of the staff would have changed, their conditions of work would certainly have changed and their workload would almost certainly have increased. What is relevant is their attitude to the service provided at the point in time at which it is measured and not comparing it with a previous and by now irrelevant state in the past. I have become increasingly convinced that an attitude survey of users is essential and a most informative part of the evaluation. By users in this context I am specifically referring to both medical and nursing staff. They are the people by whom the information is requested and who make judgements on information received from the laboratories. Their subjective opinions, if a large enough number of people are canvassed to be regarded as representative, are of great value, no matter how much the laboratory's through-put, turn-around and accuracy may be improved. If the user is not satisfied a system cannot be regarded a success.

Finally, it is essential to look at the financial aspects of the exercise. In this area I feel information is likely to be misleading. The particular project under discussion was set up, in the Department of Health's own definition, as an experimental project. This implied that the project was carrying out procedures which had not been established and therefore might not be capable of development or implementation. Given that situation, it was necessary to acquire a machine of considerable sophistication which contained many features necessary for the development of a system but not necessary for the running of such a system when established for routine use. Inevitably, because a number of different processes were interacting in an on-line system and all applications had to be defined and written from scratch, the amount of manpower required to develop the system was considerable. One can cost the total effort required, both in manpower and in financial terms, in order to achieve the final system.

If this system is to run in the one laboratory only then the unit cost is enormous. If it is acceptable to other users and can be implemented on a smaller and simpler computer then some attempt must be made to quantify this in terms of effort and money. Does one add to this a part of the original cost of development. One should, but as the Department of Health and Social Security pay for one and the National Health Service the other, it is not specified in the rules.

Then we must consider the running costs. Provision of the necessary environment and maintenance is relatively easy to cost but space and electricity cannot be costed separately with any accuracy. For example, what is a square meter of hospital worth?

One area of expenditure which is normally suspect in a computer system is that of paper. When replacing an existing manual reporting system with a computerised system it is likely that stationery costs may be reduced. It is extremely difficult to ascertain this. Under the relatively new system of functional budgeting stationery is not charged to an individual department. Therefore not only is it very difficult to obtain costs before and after but the department making any saving in this direction is not likely to see any benefit.

Having said all this, I consider it essential to attempt to determine the likely cost in capital and revenue. Nevertheless, I do not think that when various proposals are in competition the most efficient is always chosen, but rather the one which, at that point in time, has the most appeal. This appeal may be on national, local, personal or emotional grounds. There are many who believe, in the long run, that this approach pays. In such a situation evaluation can provide information which may prevent financial disaster.

There has been much criticism of evaluation over the years and many have implied that the cost of evaluation as underwritten by the Department of Health could well have been better spent on the enhancement of the systems being evaluated. It is encouraging to find a working group set up by the Department of Health reviewing evaluation and redefining the objectives(3). They also make practical and sensible suggestions as to how the information required to make informed decisions about future computer projects may be achieved. To me, one jarring note in this report, which is still in draft form is a section submitted on behalf of a group of district administrators who were asked the type of evaluation information they would require to pronounce on the suitability of a computer system for implementation in their organistion. Their document runs to several pages. The cost of collating this information accurately would consume the funds required to implement the system. Even more frightening is a footnote which reads 'The above presupposes that administrators are concerned to maximise their resources by getting more work done at the same or lesser cost thus reducing the unit cost and diverting monies to other purposes'. I find this a most dangerous philosophy. I applaud increased efficiency but it must be appreciated that there is a limit beyond which it is not possible to get more work done at the same cost. Indeed there are situations in which the service can best be met by providing the same number of patients with better care at a slightly increased cost. The idea that, if the unit cost is reduced that money must be diverted to other purposes, is, in itself, self-defeating. Few of us who have been in the National Health Service for any length of time retain sufficient of the milk of human kindness to wish to slim down our own service in order that some unrelated aspect may be enhanced.

To return to the main theme. Any new system introduced into the National Health Service should be carefully examined to determine its effectiveness and its cost. This applies very much to computer systems.

The system must be examined impartially. Wherever possible hard data should be collected. When the period of introduction of the system is so short that the work of the department concerned is unchanged there is merit in a before and after study. Financial information should be obtained but it must be realised that this information is unlikely to be comprehensive or accurate.

In the end, the decision to install a computer system depends on

1. The enthusiasm and influence on management decisions of the prospective owner.
2. The attitude of the user to the department concerned.
3. The availability of the necessary finance.

REFERENCES

1. Haematology Computer Project Evaluation: Report 1–Establishment of Baselines. Addenbrooke's Hospital Computer Services. October, 1972.
2. The Evaluation of the Haematology Laboratory Computer Project–Before Situation. Addenbrooke's Hospital Computer Services. July, 1975.
3. Draft report of a working party on computer evaluation. Computers and Research Division. Department of Health and Social Security. June, 1977.

SMAC COST EFFECTIVENESS

Adrian A.A. Clarke
Principal Work Study Officer,
Department of the Civil Service, for Northern Ireland
Belfast

INTRODUCTION

This paper outlines the cost effects on analysing blood specimens at the Royal Victoria Hospital (RVH), Belfast which were brought about by the use of a SMAC analyser instead of the previous 6/60 and 12/60 Auto-analysers and the use of a Modular I computer for processing test results instead of an IBM 1130.

The original or 'baseline' costs were studied just before the SMAC was installed, and the changed costs were studied during the technical evaluation of the new equipment as soon as that equipment was working satisfactorily.

The SMAC and Modular I Computer

The SMAC is a third generation sequential multiple analyser incorporating a built-in control computer with the capacity to perform up to 20 different analyses on each specimen of blood presented to it at a maximum specimen handling rate of 150 an hour. This compares with the combined capacity of the 6/60 and 12/60 analysers to perform up to 18 different analyses on each specimen at a maximum rate of 60 specimens an hour.

The Modular I is a newer, more powerful and more versatile computer than the IBM 1130–which was in any case reaching the end of its useful life.

BACKGROUND

In January 1974, following exploratory talks between Mr D W Neill the chief biochemist at the RVH and representatives of the Technicon Company, it was suggested that the SMA 6/60 and SMA 12/60 analysers should be 'traded-in' at 100 per cent of their original capital cost in part exchange for a new SMAC analyser.

Since the 6/60 and 12/60 had then been in use for three years and four years respectively out of nominal life spans of seven years, this arrangement would have resulted in the RVH laboratory receiving one of the latest high capacity analysers on very favourable terms.

Concurrently, the Department of Health and Social Security (DHSS) were approached regarding their possible interest in the new SMAC analyser from the viewpoint of its technical performance and workload capacity. They confirmed this interest and indicated that they would also be anxious to encourage a rigorous costing study of the operation of the SMAC with particular reference to a comparison with existing procedures. They were prepared to support a properly designed and complete evaluation, and to co-ordinate parallel evaluations in Edinburgh and Belfast.

In August 1974, after extensive discussions and favourable recommendations by the appropriate committees the proposal was approved by the Eastern Area Health and Social Services Board in Northern Ireland.

In December 1974 after some renegotiation of the financial arrangements (due to further depreciation and inflation during the discussions which had taken place throughout the year), a new SMAC was ordered for delivery in June 1975.

Co-ordinating Committee

Soon after the proposal was mooted a coordinating committee was established to organise and monitor the project. The membership was drawn from interested pathologists together with medical and administrative officers from the DHSS (London), the Scottish Home and Health Department and the DHSS (Northern Ireland).

During the committee's discussions it was decided to conduct a thorough investigation into the laboratory costs before installing the SMAC so that proper 'baseline' or reference periods costs might be established for comparison with the SMAC costs at a later stage.

Terms of Reference

As a result of this decision the Work Study Branch of the Northern Ireland Department of Health and Social Services was asked to carry out the necessary cost investigations at the Royal Victoria Hospital with the following terms of reference:

1. A comprehensive study of all costs in the manual, steroid, automation and computer sections of the Department of Biochemistry using the Cooper and Lybrand system of cost analysis in order to determine the 'baseline' unit costs of performing individual analyses.
2. Another similar study in the automation and computer sections when the SMAC and Modular I had been installed satisfactorily in order to establish the altered unit costs of performing individual analyses resulting from their introduction for comparison with the corresponding 'baseline' figures.
3. A cost study of established transport facilities associated mainly or exclusively with carrying specimens between other hospitals in the locality and the RVH.

Cooper and Lybrand System of Cost Analysis

The Cooper system of cost analysis takes into account all the costs *actually* incurred under the headings of staff, machines, materials and equipment; and then allocates them as directly as practicable according to the time and amounts of money spent on individual tests–leaving any staff costs which cannot be treated thus as 'indirect overheads' which are allocated according to the direct technician times on individual tests.

The full procedures are covered very comprehensively in a set of instructions issued by the DHSS (London). Very briefly the main sources of information are:

a. a technician time record completed daily by each technician,
b. a machine time record completed daily for each main machine,
c. financial and supplies information.

BASELINE CONDITIONS AT THE ROYAL VICTORIA HOSPITAL

So that the basis of the Royal Victoria Hospital figures would be more meaningful for other laboratories the main statistics and procedures in the biochemistry department were specified in considerable detail in the main report.

The main statistics, procedures and definitions are summarised as follows:

Specimens Tested

The main categories of specimens tested during 1974 are shown in Table 1.

Table 1 Categories of Specimens Tested in 1974

	Specimens	*Tests*
SMA 6/60	38,751	232,506
SMA 12/60	49,159	589,908
Blood sugars	38,645	38,645
Other tests	9,682	9,682
Automated total	136,237	870,741
Manual total	88,892	88,892
Steroid total	15,750	15,750
Grand total	240,879	975,383

Apart from tests which are performed very infrequently there were no significant variations between the weekly averages for 1974 and the weekly averages during the baseline study in March 1975.

Staff Employed

During the two weeks self-recording baseline study between 10th and 22nd March 1975, the total number of staff employed in the laboratory was 50. These were deployed in the following main groups:

Administration	3
Office and general	6
Phlebotomists	7
Specimen reception	4
Autoanalyser section	5
Computer section	5
Manual section	9
Steroid section	11
Total	50

Machines Used

The main machines used during baseline study were:

Autoanalyser section:	Computer section:
1 IL 413 Flame photometer	1 IBM 1130 Computing system with
1 Technicon AAI (creatinines)	1131 CPU 16k core
1 Technicon AAII (blood sugar)	1132 Line printer (80 lines/min)
1 Technicon SMA 6/60	1142 Card read/punch
1 Technicon SMA 12/60	2 129 Punch/verifying machines
	1 029 Punch

Procedures and Definitions

The agreed definitions for certain terms are best illustrated by following the normal procedures from the collection of blood through the laboratory to the issue of the report.

Sample: Normally only one syringe of blood of the appropriate capacity is taken from the patient. This is regarded as a sample.

Specimen: The phlebotomist then divides the sample into a number of 'specimens' according to the requirements of the clinician.

Request: The biochemistry department have stipulated that a separate request form should be submitted with each specimen. So we took the number of requests as being identical to the number of specimens received.

Reception and Request Handling

Normally all the specimens and requests are delivered to the specimen reception room. Here the laboratory aides sort and centrifuge the specimens for the 6/60 and 12/60 and place them in numbered positions in the appropriate test racks (or plates). The aides next select the corresponding request forms, identify them by marking them with the appropriate plate and position number and then take them to the punchcard operators. At this stage the specimens are taken for analysis: after analysis the racks, tubes and punchcards are collected by the laboratory aides who cross-check the cards produced by the punch operators against the specimens in the test racks. After this the cards are returned to the computer section and the specimens passed to the autoanalyser section where they are held for 24 hours and then scrapped.

In the costing of 6/60 and 12/60 specimens 'reception and request handling' was treated separately as the work of the laboratory aides because it comprises nearly all their work. But for other processes this 'operation' was considered as part of the technicians' work. The punching operation was covered under 'results handling' in the computer section.

Departmental Specimen Handling and Test Process

For the 6/60 and 12/60 we took these words as meaning all the routine work of the laboratory technicians. On *other processes* these two categories of work only represent a portion of the laboratory technicians' work–which also includes 'request handling' and 'results handling'.

Test: Each chemical analysis was regarded as a 'test'. Thus each specimen on the 6/60 and 12/60 represented 6 and 12 tests respectively; and in all other cases each specimen represented one test.

Results Handling and Reports

For much of the greater part of each date the 6/60 and 12/60 analysers were connected 'on-line' to the IBM 1130 computer. Thus as the tests were completed by the analysers the identity of each specimen and the results of each test were stored in the central process unit and summarised on a special 'print-out'. When one or more 'plates' had been completed and the 'print-outs' checked for errors the computer operator selected the corresponding set of punchcards and put them in the card reader. The results were then printed out onto special report forms—with one individual report for each specimen on each of the two analysers.

Apart from the results of the 6/60 and 12/60 tests the computer did not deal with routine work for any of the other biochemistry processes; but a considerable number of reports were produced for the haematology and bacteriology departments.

So the work of the computer section was regarded as 'results handling' and allocated between the Autoanalyser section, bacteriology and haematology according to the number of specimens (or reports) processed for each department.

For other biochemistry processes 'results handling' was regarded as part of the laboratory technicians' work.

Key unit: These definitions meant that the key unit in measuring 'output' was the number of specimens handled. This was central to the calculations.

RESULTS OF BASELINE STUDY

Collection, Analysis and Reporting

The total costs *per specimen* for each process in the three main sections of the laboratory were summarised in accordance with the instructions or individual 'cost summary sheets'. In this paper, however, only those results relating to the 6/60, 12/60 and the three manual tests which were due to be transferred to the SMAC are being discussed (Tables 2 and 3). The costs

Table 2 Summary of Costs per Specimen

	Pence per test				
	6/60	*12/60*	*SGPT*	*CPK*	*Triglycerides*
Specimen collection	10.08	10.08	10.10	10.10	10.10
Reception and reagent handling	14.00	14.00	10.90	10.90	10.90
Department handling and test	25.28	32.80	14.33	20.54	82.98
Results handling	4.67	4.67			
Total (excluding reagents)	54.03	61.55	35.33	41.54	103.98
Reagents	5.19	14.64	0.06	25.33	11.15
Totals	59.22	76.19	35.39	66.87	115.13

per test on the 6/60 and 12/60 were derived from synthesised costs for the reagents used on each individual test plus one sixth or one twelfth (as appropriate) of the specimen costs for staff, machines and materials.

Table 3. Costs of Individual Tests

	Pence per test	
	Reagents	*Total*
Tests on 6/60:		
Carbon dioxide	0.72	9.72
Chloride	0.81	9.81
Potassium	0.97	9.97
Sodium	0.97	9.97
Total protein	0.85	9.85
Urea	0.89	9.89
Tests on 12/60:		
Albumin	0.83	5.99
Alkaline phosphatase	0.84	6.00
Cholesterol	1.23	6.39
Lactic dehydrogenase	3.37	8.53
Serum calcium	0.72	5.88
Serum glucose	1.00	6.16
Serum inorganic phosphate	0.78	5.94
Serum urea	0.93	6.09
SGOT	0.81	5.97
Total bilirubin	1.11	6.27
Total protein	0.90	6.06
Uric acid	2.12	7.28
Manual tests:		
SGPT	0.06	35.39
CPK	25.33	66.87
Triglycerides	11.15	115.13

Transport Costs

An approximate estimate of the costs for transporting specimens from external sources was developed from the transport arrangements pertaining in the baseline period.

The main stages in the calculations were:

a. vehicle mileage and driver hours for one run,
b. vehicle costs and driver costs for one run using AA costs per mile and hospital costs per hour worked,
c. total costs per run from each source,
d. total costs per annum for each source using the frequencies shown,
e. an arbitrary estimate of the proportions of the total annual costs chargeable to biochemistry specimens from each source.

Centralising Biochemistry Facilities

The interest in transporting specimens from external sources is largely associated with the centralisation issue. By calculating the existing total costs of analysing biochemistry specimens at external units and then recalculating the cost of testing them at the RVH plus transport costs

COST STUDY OF SMAC AND MODULAR I

In July 1976 when the SMAC was working satisfactorily, a second study was carried out to ascertain the cost effects of using the SMAC in place of the 6/60 and 12/60 analysers and the Modular I computer in place of the IBM 1130.

The study was confined to the automated and computer sections, i.e. the sections where changes had occurred. Apart from these two machine changes, most other aspects of the situation, with the exception of reagent costs and the number of specimens tested annually, were similar to those in the baseline study. Furthermore, in order to counteract the effects of inflation and make the cost comparisons more valid it was assumed that there had been no change in salaries or the capital cost of machines between the two studies.

In the case of reagent costs, it was not possible to discount for the effects of inflation, but these form a small proportion of the total cost.

In the case of the number of SMAC specimens tested annually no figures were available so a figure of 123,552 specimens was assumed on the basis of experience at the Royal Infirmary, Edinburgh. Nonetheless, the actual study values for 'man-minutes per specimen' and 'machine-minutes per specimen' were used in the calculations. These values were not adjusted on the basis of the assumed annual output.

Results of the SMAC Study

The total costs *per specimen* were calculated from the results of the study in accordance with the instructions issued (Tables 4 and 5). The costs per

Table 4. Costs per Specimen on SMAC: 19 Channels using RVH Reagents

	Pence per specimen					
		Staff		*Machine*		
	Material	*Direct*	*Indirect*	*Running*	*Depreciation*	*Totals*
Specimen collection	8.40		1.53			9.93
Reception and reagent handling			12.33			12.33
Department handling and test		8.26	1.55	10.83	12.75	33.39
Results handling		2.35	0.56	3.24	2.41	8.56
Totals (excluding reagents)						64.21
Reagents (19 tests)	8.20					8.20
Totals	16.60	10.61	15.97	14.07	15.16	72.41

test on the SMAC were derived from synthesised costs for the reagents used on each individual test (RVH and bought reagents) plus *one nineteenth* of the specimen costs for staff, machines, and materials.

Table 5. Costs of Individual Tests on SMAC:
19 Channels using RVH and Bought Reagents

	Pence per test						
	RVH reagents		*Bought reagents*		*Total costs on:*		
	Reagents	*Total*	*Reagents*	*Total*	*6/60*	*12/60*	*Manual*
Albumin	0.34	3.72	0.52	3.90		5.99	
Alkaline phosphatase	0.32	3.70	1.91	5.29		6.00	
Bilirubin total	0.30	3.68	0.87	4.25		6.27	
Calcium	0.26	3.64	0.36	3.74		5.88	
Carbon dioxide	0.42	3.80	0.38	3.76	9.72		
Chloride	0.25	3.63	0.36	3.74	9.81		
Cholesterol	0.41	3.79	0.83	4.21		6.39	
Creatinine	0.25	3.63	0.35	3.73		(5.94)	
CPK	1.60	4.98	9.38	12.76			66.87
Glucose	0.28	3.66	2.59	5.96		6.16	
Lactic dehydrogenase	0.58	3.96	2.01	5.39		8.53	
Potassium	0.31	3.69	0.65	4.03	9.97		
Serum iron	0.31	3.69	0.42	3.80			
SGOT	0.43	3.81	1.27	4.65		5.97	
SGPT	0.93	4.31	1.32	4.70			35.39
Sodium	0.31	3.69	0.41	3.79	9.97		
Total protein	0.25	3.63	0.37	3.75	9.85	6.06	
Urea nitrogen	0.25	3.63	0.39	3.77	9.89	6.09	
Uric acid	0.39	3.77	1.23	4.61		7.28	
Costs per specimen	8.20	72.41	25.59	89.80	59.21	76.56	–

The changes in the tests and effective results between baseline and SMAC periods are as follows:

a. The 6/60 and 12/60 together performed 18 actual tests, but the total protein and serum urea tests were duplicated; thus only 16 effective results were obtained.

b. The SMAC performs 15 of these effective results–i.e. all the tests listed in Table 5 except Creatinine, CPK, SGPT and serum iron.

c. The sixteenth test (serum inorganic phosphate) has been replaced by the creatinine test. For simplicity these two tests have been regarded as equivalent in subsequent cost comparisons.

d. In addition SMAC does SGPT, CPK and triglyceride tests which were done manually but now triglycerides are done separately in batches as outlined later.

e. The SMAC also does serum iron tests which were done in haematology in baseline period.

Testing Triglyceride Specimens on SMAC

Owing to the importance of taking triglyceride samples from fasting patients and the inevitable waste of expensive reagents and of patient specimens if this is not done, it was decided to do these tests in two separate batches per week where reagents were run through all 20 channels but only the triglyceride and cholesterol results were used. From a costing point of view, however, it has been argued that it would be misleading to split the specimen costs between these two tests; so all the costs (including those for the 20 reagents) were debited to triglycerides (Tables 6 and 7).

Table 6. Costs of Triglyceride Specimens on SMAC using RVH Reagents

	Pence per specimen					
		Staff		Machine		
	Material	Direct	Indirect	Running	Depreciation	Totals
Specimen collection	8.40		1.53			9.93
Reception and reagent handling			12.33			12.33
Department handling and test		8.26	1.55	10.83	12.75	33.39
Results handing		2.35	0.56	3.24	2.41	8.56
Total (excluding reagents)						64.21
Reagents	16.40					
Totals	24.80	10.61	15.97	14.07	15.16	80.81

Table 7. Costs per Triglyceride Test or Specimen

	Pence per test or specimen		
Type of Reagent	Common	Reagents	Totals
RVH	64.21	16.40	80.81
Bought	64.21	32.00	96.21

COMPARISON OF BASELINE AND SMAC COSTS

The most objective method of comparison seemed to be to take the actual cost (in 1974) of the three 'manual' tests due for transfer to the SMAC plus the actual cost of the specimens tested on the 6/60 and 12/60 in 1974 (the last year not affected by SMAC) and compare it with what it would have cost to obtain the same results using SMAC without making any allowance or credit for the extra results from SMAC until later.

The first step in doing this was to determine the equivalent number of specimens which would be needed. The number of SMAC specimens must allow for those cases where two specimens per patient were required in the pre-SMAC situation but where one specimen will suffice for SMAC.

A special survey showed that 29 per cent of the 6/60 specimens were duplicated on the 12/60. So the same results (plus others) could be obtained from SMAC with 12.75 per cent or 11238 less specimens than the combined total tested on the 6/60 and 12/60 in 1974.

In the case of the three 'manual' tests there was no data on duplication so it was assumed that the full number of specimens were required for SMAC. A similar assumption was made in the case of serum iron tests done in haematology in the baseline period.

The costs of the baseline and SMAC periods were then calculated by multiplying the numbers of specimens and the equivalent number of specimens by the appropriate specimen costs.

Various cases or combinations of the different reagents and computers were considered. These are summarised in Table 8.

Table 8. Comparative Costs (£s) of 1974 Output: SMAC Baseline[b]

	SMAC	*Baseline* [b]	*Difference*
RVH reagents: Modular I			
Analysis[a]	50026	53419	- 3393
Reagents	7685	10472	- 2787
			- 6180
Results	7695	4105	+ 3590
Overall	65406	67996	- 2590
Bought reagents: Modular I			
Analysis[a]	50026	53419	- 3393
Reagents	23249	31744	- 8495
			-11888
Results	7695	4105	+ 3590
Overall	80970	89268	- 8298
RVH reagents: IBM 1130			
Analysis[a]	50026	53419	- 3393
Reagents	7685	10472	- 2787
			- 6180
Results	4198[c]	4105	+ 93
Overall	61909	67996	- 6087
Bought reagents: IBM 1130			
Analysis[a]	50026	53419	- 3393
Reagents	23249	31744	- 8495
			-11888
Results	4198[c]	4105	+ 93
Overall	77473	89268	-11795

a Includes collection and reception costs
b Baseline: 6/60, 12/60, IBM 1130, three manual tests
c Hypothetical: Using IBM 1130 with SMAC

National Cost Benefits of SMAC

As all the cases considered above showed substantial tangible benefits resulting from the use of SMAC and its reagents, it was then possible to attribute the notional (but entirely additional) cost benefits of the extra results provided by SMAC.

This was done by calculating what *additional* sum it would have cost in 1974 to provide the full range of SMAC results for all the specimens tested on the 6/60 and 12/60.

Five per cent of the resultant sum was arbitrarily considered to be the notional cost benefit of the extra SMAC results. For RVH reagents and bought reagents the figures were £7,449 and £10,138 per annum respectively.

Size of Laboratory Which Would Justify the Purchase of a SMAC

A final series of calculations showed that the purchase of a SMAC would be justified in laboratories with an annual workload of not more than 100,000 specimens and arguably as low as 80,000 specimens (i.e. specimens which would normally be analysed on SMAC).

SUMMARY OF MAIN CONCLUSIONS REGARDING COST BENEFITS OF SMAC

At the 1974 workload the tangible annual cost benefits of using a SMAC with bought reagents (as compared with the old analysers and bought reagents) would be £11,888.

If reagents made at the RVH were used the tangible annual cost benefits would be increased by £15,564 to a total of £27,452 (as compared with the old analysers and bought reagents).

In addition to the tangible cost benefits there would be intangible benefits of £10,138.

Since the SMAC produces large amounts of data and inevitably requires better external data processing equipment, the savings quoted above would be offset by £3,590 due to the higher costs of the Modular I computer (as compared with the IBM 1130).

The purchase of a SMAC analyser would be justified in laboratories with an annual workload of 100,000 specimens (i.e. specimens which would normally be analysed on SMAC).

The cost per specimen for a SMAC analyser running on 19 channels at a daily workload of 450 specimens plus 126 at weekends using RVH reagents would be 72.41 pence.

ACKNOWLEDGEMENTS

The work study officers who conducted the cost evaluations are very much indebted to many professional and administrative officers from the Health and Civil Services in England, Scotland and Northern Ireland. But especially so to Mr D W Neill, Chief Biochemist at the Royal Victoria Hospital, Belfast and his staff.

SHARED FACILITIES–A STRATEGY FOR SURVIVAL

F. Siemaszko BSc
Principal Scientific Officer, Bristol Health District

INTRODUCTION

The remarks in this article apply to small and medium computer installations in Health Service clinical laboratories and offer some suggestions for alternative uses of such computers. The need for alternative applications may be judged by considering the history of a typical small installation as it nears the end of its development of a primary laboratory system. The composite picture which follows in the next few paragraphs is based on a questionnaire circulated to clinical laboratories in 1976 and each reader will of course judge whether it applies in his case or not.

It is probable that our representative laboratory decided to develop a system largely to its own specification and it is equally probable that the first attempt had underestimated the diversity and time-interdependence of laboratory processes. A revised system would have generally been attempted to remove the principal causes of inflexibility, bottlenecks, and plain omissions of the first design. This would have been achieved not only by a revision of software but also by accommodation on the part of the laboratory as it learned to make better use of the computer based functions. Eventually a system that gave satisfaction to our particular laboratory would have evolved.

In the context of routine laboratory operation the computer would now appear to be working well but, despite all, some nagging doubts might remain.

Firstly, as has already been noted, most UK systems are self designed so it is likely that associated with the laboratory there would now be one or more seasoned specialists capable of both programming and of system design. Without new computing development they would be under-employed but their services, even with a seemingly completed laboratory system, could not be entirely dispensed with and some means of ensuring continued programming support would have to be found.

On the question of cost effectiveness the laboratory would by now have realised that in assessing their system overall it was difficult to ask the right questions and often impossible to answer them, the cardinal difficulty being to find like to compare with like. The one certainty would be that the computer and its associated staff were expensive and that more use, sensible if possible, should be made of both.

ALTERNATIVES

A number of broad alternatives are open to any installation which is in the state described above.

One possibility would be some attempt to intensify the automation of the laboratory. An example would be the incorporation of machine-readable labels to provide automatic and positive sample identity. Opinions differ as to the extent of automation that is desirable but insofar that such schemes use microprocessors, or other means, to perform their new functions largely in isolation from the central, clerical, laboratory computer they could make the laboratory's life easier. It must also be remarked in passing that persistent 'over kill' in terms of hardware could in the long term bring the laboratory computing into disrepute. Because this is an article about cost effectiveness we will assume that, at least for the time being, our impoverished laboratory can not afford the extra hardware and we will not discuss this option further.

Another possibility would be some development along the lines of the much heralded 'stage II'. Proposals of this kind imply a greater involvement of the computer in clinical matters. A common feature of such projects is that ultimately they are based on an analysis of more general medical information in conjunction with laboratory test results and therefore need the active, and time-consuming, support of clinicians. Apart from a variety of other considerations this need for support can effectively stop 'stage II' developments in all but the larger installations where adequate medical staff, willing to pursue numerical medical research, is available.

The remaining possibility is to provide a general service to other departments within the hospital. This approach has had very few advocates, in writing at least, so that a case for it will be made out here. At first sight it may appear inappropriate for inclusion in a book on laboratory computing but it is a means of keeping the computer team in being and, as will be seen later, other benefits can accrue to the laboratory. In the prevailing economic situation it may be an option that more laboratories will have to consider.

SUITABILITY OF COMPUTER INSTALLATION

If the laboratory had incorporated some means of concentrating test data in time before passing them to the clerical computer (rather than operating analysers direct in real-time, on-line, to the main processor) then there would be periods during each day when the whole of the system could be given over to other tasks. A typical pathology computer, as opposed to say a radiotherapy installation, will be equipped with a fast printer and will have been bought with an emphasis on good filing facilities so the hardware is suited to applications of a commercial nature. If the laboratory installation is still equipped with old-fashioned cards then the new users would have the luxury of preparing data off-line in their own time, otherwise it would have to be on the basis of time-sharing or background use of the computer. (The data entering function should in any case be kept separate from the data processing function, as explained later, in the kind of application that we are considering.)

In every hospital group there are many minor (and major) tasks that could benefit from the use of a computer but many of them are not suitable for running on a regional mainframe. This could be because they require too fast a turn around time or else because they may need some user intervention to run properly. In many cases the jobs are simply not automated because they appear too small to make a special application to the regional organisation.

Pathology computers are generally at a convenient central site with respect to the hospitals they serve. Their staffs are used to judging man-machine interaction and to coping with tasks that need a tight schedule, and in addition, from the nature of their work, they have some understanding of hospital organisation. It is suggested that many hospital management functions are ideally performed on pathology computers.

SOME POSSIBLE ADDITIONAL USES

Before suggesting how to tackle projects with limited computer and staff resources a list of some possible and simple applications is given. They have all been in routine use on our pathology installation for some time and so are not fanciful. Since no external department is compelled to use our services, and each must provide its own operating staff, and most have in fact contributed to the hardware, it is presumed that they find the computer effective. (A number of other projects, which were found to give a poor return on programming and operating effort are not included in the list but will be alluded to in the next section.)

a. Admissions, transfers and discharges of inpatients: maintain an alphabetical master file which is updated once a day. Print various lists (any medical records department will say what is necessary) and myriad address labels. The index of performance here is the number of times that a patient's particulars are sorted and printed under genuinely useful headings in return for the labour of punching or typing them once.

b. Outpatient clinic appointment lists: similar to the above. Because no master file need be maintained this is a simpler task to write and to operate and might be a suitable project to start with. If the lists are prepared some days before the appointments are due advantage can be taken of the varying daily pathology work loads to do the printing runs when there is less pressure on computer time.

(Many hospitals have, of course, developed comprehensive computer-based patient record systems by prime intent. What is suggested here is that by working in batch mode and omitting non-essential niceties a records system, which is labour saving, is possible with the use of spare capacity on what was previously considered a dedicated machine.)

c. Incidental benefits which can accrue from the labour of preparing data for a and b above are:

i. automatic extraction of some of the mandatory statistics required by the Department of Health and Social Security.

ii. printing of lists to allow the daily pulling of case notes by one non-recursive walk through the shelf area. (The notes are usually stored by order of the terminal digit of the hospital registration number. In our case a trivial program, taking about one minute to run, is said to save several hours of wandering about each day.)

iii. sometimes patients are moved after a test request form is sent to the laboratory and an efficiently maintained master list helps the result to reach the right destination. Various means of establishing an automatic interplay between the laboratory and hospital files might be tempting but, initially at any rate, could sap entirely the resources of some of the small installations.

d. Maintain a commodity price list, which, in conjunction with a file of recipes, allows pricing of canteen meals as required. Catering managers are not given a free hand in pricing but must follow a formula which makes the permitted charge a function of current food costs. In times of inflation the need for recalculation is a constant headache which occupies an inordinate amount of their time.

e. Use the price list of d above to monitor inpatient feeding costs by translating numbers of menus served (or more simply quantities of commodities entering kitchens) into costs on say a weekly basis. A common complaint of similar schemes run on regional computers is that, because of the many remote steps involved, by the time the information becomes available to the catering manager it is only of historic interest and cannot be acted on.

f. From a file itemising all planned maintenance jobs to be undertaken by the works and building department within the hospital print a label for each one that is due this week. This is a very simple program that for a few minutes of computer time saves a lot of clerical effort. A slightly more ambitious scheme would balance the standard hours for each trade for each week.

g. Although it may be difficult to get clinicians involved in computer based pathology research there are usually many problems within their immediate ambience which they feel could profit by being better understood as a result of some kind of survey. A common feature of such surveys is the desire to accumulate fairly massive patient data, occasionally to test a specific hypothesis but more usually in a general 'retrospective' mode, and then to form various groupings or contingency tables as prima facie evidence for an underlying process.

The evidence being sought is generally like the proverbial two grains of wheat in a bushel of chaff and the ratio of negative to positive findings is very high. The method is labour intensive, nevertheless it may be the only means of making progress in clinical situations where experimentation is not appropriate.

The expressed needs of various groups requesting this kind of service are so similar that it is possible to write a general, parameter-drive package, that actually works for different groups without programming intervention. Each user specifies his own data format to the computer and provides the operating labour.

PROVIDING A SERVICE WITH LIMITED RESOURCES

Some practices which have been found useful in providing a fringe or background service are given. As with most pragmatically derived rules exceptions will be found and what follows is not meant to be an ideal way of coping without exception in every case.

a. In assessing a proposed project determine at the outset what the benefits are likely to be. If these cannot be given in clear and simple terms, and imponderables have to be resorted to, there is no need to start the project. An advantage which the fringe service has over clinical laboratory computing proper is that no one is compelled to use the computer, and reciprocally, since in a sense the computer belongs to the pathology laboratory, the external user cannot insist on the exact form of the service to be provided for him. One can be highly selective in one's choice of projects and this freedom should be exploited.

b. The department for which the work is being done must provide its own data preparation and computer operating staff. This has several advantages. One is the implication that no extra staffing around the computer will be required. Another is that people preparing data in a field with which they are familiar are more likely to spot errors than full-time operators; and if the actual data preparation is only one part of their duties they are less likely to get bored and resign. The cardinal advantage, however, is the effect that such an arrangement has on the overall design of any given project.

The cost of software development is frequently mentioned but it must be realised that in any system which becomes routine the operating costs must eventually outweigh development costs by several orders. When the potential user realises the full extent of his commitment it will steady his mind wonderfully and a respecification of the requirements to deal with essentials only is likely. If the weight of data preparation still seems too great for the expected benefits then, as already stressed, there is no need to embark on the project. If an apparently free service were to be provided without any effort being necessary on the part of the recipient of the service, then he would of course accept it uncritical of its cost.

c. Because the computer is to be run by non-specialist and possibly changing staff it is particularly important to make the software modular and to divide each task into strictly segmented operating stages. What is meant here is illustrated by an example. Suppose that for a medical records application, patient identity cards for new admissions are being read in order to merge them into a master file of inpatients. The reading and syntax checking stage should be made quite separate from the subsequent merging stage. Once the stack of cards is read the user has the option of correcting and rereading the cards or of going on, directly. He might have spotted errors not detectable by the computer, or alternatively because of time pressure or other exigencies, he may decide to go on although minor syntax errors have been indicated. In a similar way the subsequent stages should be segmented.

In terms of sheer computer performance the segmented approach is not the quickest way of running a job and, indeed, for a clinical laboratory application it may be too fussy. For a background service, however, it has the following advantages:

i. the fringe task can be interrupted to allow a more pressing job to be run;
ii. following an operating error relatively unpractised operators can more often get out of trouble themselves without calling on a computer specialist;
iii. in the event of a major shemozzle, recovery by the computer staff is easier since the breakdown is readily traced to a given stage.

d. In a way that complements the considerations in c above it is necessary to put as much onus for decisions on the user as possible by providing a suite of relatively simple computer functions. If, on a given occasion, a computer run does not give a satisfactory outcome it is best if the fault is seen to be the users and not the computers. One is not being cynical. No department should feel that it is absolved from responsibility simply because a part of its service is automated.

If control is taken away from the user he is given in exchange the delicious facility of sitting back and blaming it all on the computer.

In our context the true purpose of the computer is to cope with the drudgery in a given job. It should be apparent that the first thing to get right is the man-machine interface and to worry about the internal computer technical problems second.

e. A fringe service implies doing some of the work in the evening and working predominantly in batch mode. It is important to provide file editors to cope with the inevitable data errors that will arise and some of these will have to be in reactive or transactions mode.

Two applications attempted on our installation which did not adhere to the above practices proved to be unworkable and have been dropped. In one case the project was vastly over designed for its workload and overall less effort would have been needed to do it manually. In another the user thought that he would gain access by means of the computer to some information normally denied him but did not declare this as his main reason for wanting the service. Three man-months of programming effort was wasted. It may be significant that in each of these two cases senior medical staff were involved who were in a position not to accept our ground rules.

CONCLUSION

As a clinical laboratory computer reaches the final stages of its prime development the principle of diminishing returns on programming effort applies. This is because one has to interfere with a live system in order to continue development and also possibly because some of the later additions are of uncertain value. In such circumstances it could be more effective to develop a fringe service.

The departments for whom such a service is provided could not normally justify their own computing facilities but in general are able to make a contribution towards the hardware. If the facilities which are centred on the laboratory are built up by this means to include an independent means of development then eventually the computer service to the laboratory could prove to be far better than would have resulted from a policy of strict independence.

In our own case the policy of sharing facilities has given the laboratory adequate hardware for backup and justifies the continued employment of three computer specialists who are to a large extent at the disposal of the laboratory so that new pathology development is taking place in better circumstances than would have been otherwise possible.

Part 4

EXPERIENCE OF INPUT/OUTPUT DEVICES IN LABORATORY SYSTEMS

VISUAL DISPLAY UNITS AS INPUT/OUTPUT DEVICES FOR A COMPUTER IN A CLINICAL CHEMISTRY LABORATORY

A M Bold
Consultant Chemical Pathologist,
Clinical Chemistry Department,
Queen Elizabeth Medical Centre, Birmingham

Visual display units (VDUs) may act as an output device for a computer, and in conjunction with a keyboard may also serve to send information to a computer. Their general reliability, simplicity of use and flexibility are making their use increasingly popular. In our department we have had considerable experience with Cossor VDUs, and more recently with Digital decscope VDUs. This review is not concerned with the technicalities, but describes experience from the user's viewpoint, and illustrates some uses of VDUs in a clinical chemistry laboratory.

VDUs AND THE REAL-TIME COMPUTER SYSTEM AT THE QUEEN ELIZABETH HOSPITAL

Prior to 1975, the clinical chemistry laboratory at the Queen Elizabeth Hospital, Birmingham, employed an IBM 870 punchcard system in conjunction with an IBM 1130 Computer for processing requests, and generating reports (1, 2). From 1973 the system was modified and considerably improved by the production of biochemical profile reports by the computer directly from tapes generated by the SMA 12/60. The bacteriology department also used a punchcard system and an IBM 1440 Computer for handling requests and reporting (3). The haematology department used punched cards for booking in requests and producing strip reports.

In 1972 a Univac 418/III Computer System was installed in the hospital as the basis of an experimental project in real-time hospital computing. The general principle underlying the project was that some 48 VDUs (and a smaller number of teletypes) situated in all wards and in other clinical areas, and in all laboratories, would act as the principal input–output devices for a hospital-wide computer system. Handling of all laboratory requests and reports was a requirement of the system. A comprehensive patient administration system (registration, ward bed states, patient information system, etc.) was the fundamental basis for all other applications, which apart from all laboratory reporting included a nursing record system and drug information and prescribing system, and various applications in batch processing mode.

The introduction of the real-time system for laboratories in 1975 enabled us to compare the advantages and disadvantages of a system based on punchcards with a system based on VDUs. The anticipated advantages of VDUs are summaried as follows:

1. Improved legibility of requests
2. Reduced errors due to defaced or damaged punchcards

3. Increased speed of entry of requests and reporting of results, resulting in reduced need for data processing staff
4. Improved conditions for data processing staff by reduction of noise levels
5. Reduced delay in laboratory reports reaching the ward
6. Reduced error rate by improved checking procedures

However, any assessment of VDUs is complicated by the problem of distinguishing their intrinsic properties from the advantages and limitations imposed by the whole computer system. While this is not the place for a detailed description of the laboratory systems, Figures 1 and 2 summarise the punchcard and VDU systems respectively. Since biochemical profile requests, analysed on a Technicon SMA 12/60, and certain other automated tests are processed separately in both systems, they are excluded.

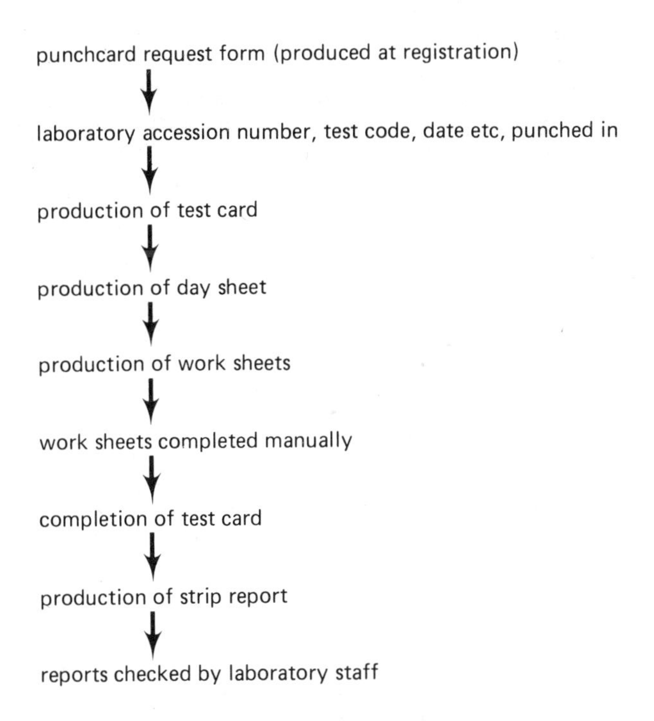

Figure 1. Outline of punchcard system (IBM 870)

In general, items 1 to 4 in the above list are due to differences inherent in the two systems, whereas 5 and 6 were facilitated by the use of VDUs but could have been achieved at least in part by alternative methods.

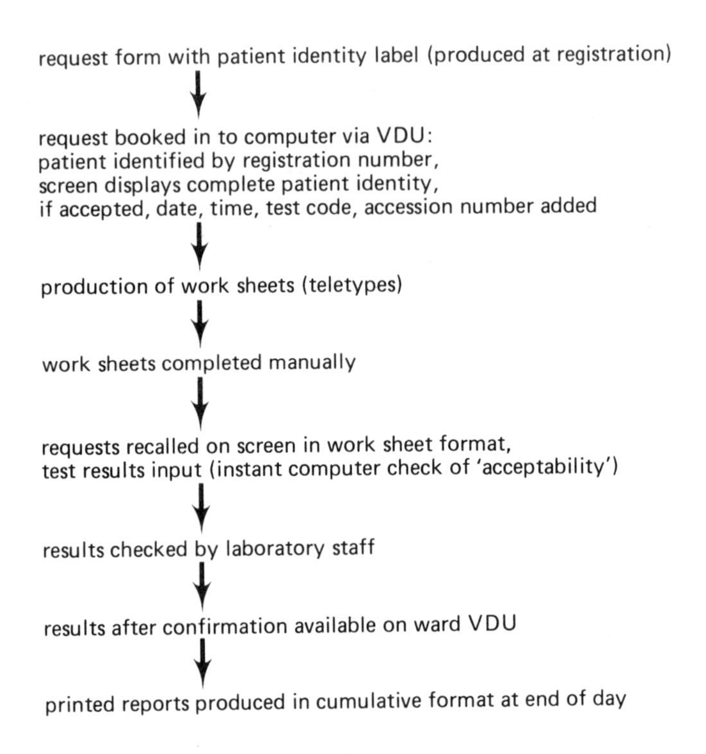

Figure 2. Outline of 'real-time' laboratory system

In our experience, the printed interpretation of punchcard data was often of indifferent quality, though readily decipherable in most cases. However, infrequently, it was impossible to read the patient's identification data with certainty in the laboratory, and on the ward occasional serious errors were caused due to confusion between patients with similar names. The use of request cards using labels preprinted at registration with full patient identification data very easily readable, in place of cards similarly prepunched at registration with patient identification data, was fundamental to the new system. Under the new system, routine requests (i.e. for patients already registered and therefore on the main hospital computer files) were entered via a VDU and identified by the patient's unique registration number of the type G123456/7 which includes a terminal check digit. There were fears that for all the theoretical advantages of the real-time system, this procedure, replacing machine readable identification data by a number manually typed into a VDU, would lead to increased errors. Although no detailed lengthy evaluation of this has been carried out, our experience over two years has been that this has not been so.

In spite of continued education of medical and nursing staff over a period of seven years, occasional defaced punchcards continued to be received, for example with the printed interpretation of the name or ward crossed out and changed by pen. It seemed difficult to get across to everyone that card readers read the holes, and lack the sense to look at the writing! Also with the punched card system, in spite of the availability of an efficient system for replenishing punchcards for patients, a considerable proportion, as little as 5 per cent for some wards but around 50 per cent for others, of requests were made on blank cards, requiring laboratory staff to generate manually a new card by punching the identification data.

It is invalid to compare punching rates on punch machines with rates on VDUs, since the actual requesting and reporting systems were different. Nevertheless, the VDU system is agreed by all staff to be substantially faster than the punchcard system, and as proof of this, in spite of an increasing workload in the pathology laboratories, the number of data processing operators has been reduced by one whole time and one part time operator since the introduction of the real-time system, and could be reduced still further were it not for the need to maintain a large enough pool of staff to cope with annual leave and sickness. No one who has worked with a punchcard system will dispute that the quiet operation of VDUs is a significant advance.

INTRODUCTION OF VDUs

The new pathology request entry/reporting systems were introduced in stages, bacteriology in March 1975, clinical chemistry in November 1975, part of haematology in 1977, the remainder to be introduced in 1978. For each laboratory system this involved training the data processing staff in the new system, as well as continuing to use the old system, and after basic training, both systems were run in parallel for a period of a month. This was essential to correct any program faults. For a laboratory activity which is a vital part of the service, it is also essential that operators become thoroughly familiar with the system so that its routine introduction can proceed smoothly. Further, it involves the operators in the system, and their comments and criticisms were carefully considered. Some specific comments on details were received. For example, the number block on the keyboard was in accordance with practice on adding machines, but upside down compared with punch machines, and new blocks were necessary; although the VDU screens were in principle very legible, it was found important for the operators to be sitting at an individually optimum height, and chairs adjustable for height were obtained. This necessitated a different design of table so that all operators could get their knees comfortably under the table! Attention to such apparently trivial detail is important.

ACCEPTABILITY OF VDUs

The operators, all with previous experience of the punchcard system, unanimously agreed that the new system was better than the old system. Speed of operation, simplicity, silence and doubless novelty were all appreciated. The easy legibility and restful colour (green) of the data on the screen were liked. The VDUs have proved extremely robust with only occasional breakdowns (about once/twice a year between the six videos in the laboratory block, all due to one VDU).

However, not all of this was gain. Over the course of the last two years, as the novelty has worn off, it has become apparent that 'job satisfaction' has decreased. Since data entry, whether via punch machine or VDU is inevitably repetitive, a certain amount of boredom is to be expected. But the reasons given for reduced job satisfaction are enlightening. Our punch machines, card readers and slave typewriters were well worn by 1975, and breakdowns were a constant source of anxiety, requiring considerable skill to operate accurately, and one had to cajole them to perform satisfactorily. Subjective assessment is always prone to error, but the need to continue with the punchcard system for haematology, used by the same operators, permits a comparison over two years of parallel use. The operators find the VDU system 'too simple', so that there is little sense of achievement in operating efficiently, whereas with the punch machines successful operation was felt as due to the operators' skill. Further, with the punched card system constant attention was required to ensure accurate punching. With the VDU system there is some control by the computer, e.g. for any given test, limits are set (for serum sodium this might be say 100 to 180) and values outside this are instantly rejected by the computer though this rejection can be overridden by the operator. This step was provided to attempt to reduce, e.g. decimal errors or errors due to inserting data in the wrong column. In the event, it is resented, albeit mildly, by the operators since they feel that the computer has taken over some of their responsibility for accurate reporting. Eason (4) has drawn attention to this problem, advising that for those whose work was primarily linked to a computer system (e.g. data processing operators) 'the system' should provide job satisfaction. This is perhaps easier said than done, but I fully support his contention.

SOME USES OF VDUs IN A CLINICAL CHEMISTRY LABORATORY

There are, in general, three types of VDUs:

1. Alphanumeric display only
2. Alphanumeric with limited capability for graphical display
3. Most sophisticated (and expensive) type, with additionally a full range of graphical displays with and without colour.

The laboratory requirements, and money available, will determine which type is obtained. Our Cossor VDUs are of type 1, the Digital decscope VDUs of type 2.

The main categories of use of a VDU are:

1. As input devices

In our laboratory as a whole, a variety of input devices are used, including VDUs, punchcards and paper tape. Input of data via a VDU is thus complementary to other methods, and VDUs are particularly useful for urgent tests, tests performed in relatively small batches and in more general context, data not readily adapted to other methods (e.g. descriptive reports as in histopathology or microbiology).

They could be used for input of data in large volumes, but are obviously slower, more tedious and more prone to error for such procedures than say paper tape generated by a high capacity analyser or an isotopic counter; entry of data via VDU is also expensive in computer time unless there is a multiasking capability.

2. As an interactive device for 'talking to the computer'

i. Checking

Quality Control of analytical steps in laboratories has been widely introduced in recent years with undoubted improvement in reliability. However, Quality Control of both preanalytical steps is less satisfactory. Yet these steps include correct identification of pathology specimens, (and maintaining identity through all analytical procedures) and transcription and other clerical procedures, where any failure can cause potentially enormous errors (5). Before reports are available to clinicians we employ a double checking procedure for all reports where data is entered via VDUs. On VDUs, members of the technical staff check against original completed work sheets that the data has been correctly entered to the computer

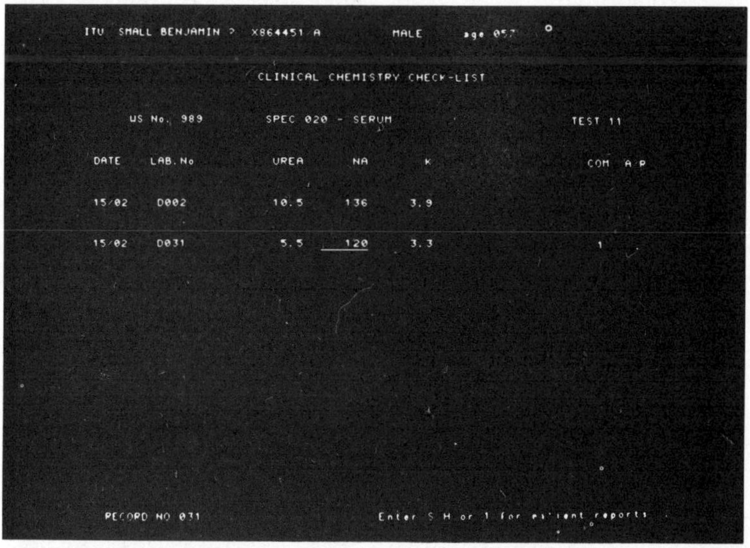

ITU SMALL BENJAMIN ? X864451 A MALE age 057

CLINICAL CHEMISTRY CHECK-LIST

US No. 989 SPEC 020 - SERUM TEST 11

DATE	LAB. No	UREA	NA	K	COM	A P
15/02	D002	10.5	136	3.9		
15/02	D031	5.5	120	3.3	1	

RECORD NO 031

Figure 3. Check list on Cossor VDU screen*

*Figures 3 to 6 are photographs of VDU screens which are technically difficult to photograph and do not do justice to the legibility of the data on the screens.

files. Once this is confirmed most results are then available on ward VDUs, and later in printed format. Any results which fall outside pre-determined limits (arbitrarily decided by senior laboratory staff because results outside these limits are either unusual or interesting) or, which differ markedly from any results within the previous ten days (i.e. by more than an arbitrarily defined amount) are held back by the computer for final acceptance, investigation or rejection by a senior member of the laboratory staff (Figure 3)). The limits can be readily changed, and obviously this checking stage can be made as sensitive or insensitive as required, and results held back can be followed up with whatever degree of enthusiasm laboratory staff wish to put into it. We have found it useful as a means of drawing the attention of senior laboratory staff to 'interesting' patients, thus helping clinical involvement of laboratory staff, without the need for the tedium of daily sifting personally through hundreds of laboratory reports. It has also revealed occasional slips where two patients' sera have been transposed, spotted because unlikely changes had occurred between repeated tests on a patient. Such a checking step cannot possibly pick up all clerical and other errors, and unless inordinate effort is expended in following up discrepant results, errors may well still be undetected. It could be argued that gross discrepances will be noticed by clinical staff, but it surely helps the laboratories' credibility if as many errors as possible can be intercepted, and also draws attention of the laboratory staff to any weaknesses in organisation.

ii. Recently we have introduced a system for monitoring and calculating serum thyroxine assays by radioimmunoassay, using a Digital

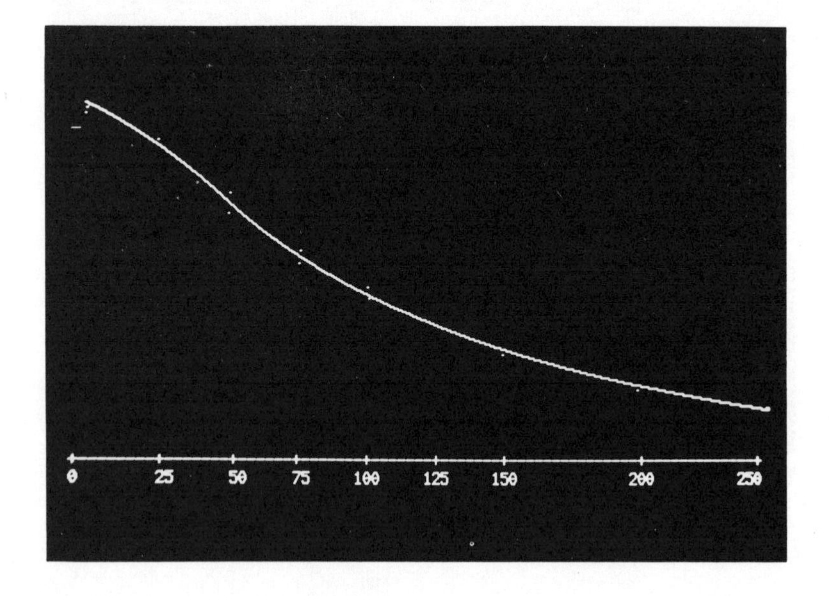

Figure 4. Thyroxine radioimmunoassay standard curve on a Digital dec-scope VDU

decscope VDU on-line to a PDP 11 40 Computer. A tape providing patient identification data is matched with a tape from a gamma counter, and input to the computer via a tape reader. The biochemist or technician in charge of the assay converses with the computer via the VDU, providing additional information, e.g. any changes in the order of tests. The data from the standard curve is then processed and displayed on the VDU (Figure 4), permitting the operator to accept it or perhaps reject one point on the curve, in which case the curve can be redrawn eliminating the data from that point. This is a good example of the use of a VDU in interactive mode, enabling the operator to monitor the process and retain control.

3. **Transient data-Progress reports for wards**

Although, like many clinical chemistry laboratories, we complete on the same day the analysis and reporting of the majority of requests received by midday, there are still many queries, e.g. whether a request and specimen has been received. Progress reports can now be obtained on VDUs, in the wards, so that once a request has been booked in, this fact is noted. This enables ward staff via a VDU to follow the progress of a test through the laboratory, and is particularly useful for tests, as in microbiology, where the laboratory investigation may proceed in stages and the final report be delayed. The information contained in the progress report is of transient value, and is not duplicated in hard copy form.

4. **Graphical display–quality control data**

For years our Department has adopted a variety of quality control procedures, including assay of pooled serum and commercial sera, calculation of daily means and running daily means from analysis of all patients' sera with values falling within arbitrary limits. Such data has been plotted manually on charts to provide a visual aid to any trends. This data is now processed by a computer and presented graphically on VDUs (Figure 5) with considerable saving in time.

5. **An alternative to hard copy**

1. Laboratory reports–in our laboratory, reports are of two types:
a. routine, covering the 35 most commonly requested tests which are presented in cumulative format in columns, covered by three VDU screens, available as soon as the reporting and checking procedures have been completed (Figure 6).
b. miscellaneous tests, presented in chronological order. Statistical interpretation is included for the routine tests together with an assessment of 95 per cent analytical confidence limits. All abnormal miscellaneous tests are underlined, and there is also the capability of adding a comment. Hard copy of reports in the same format as on the VDUs is provided, for routine tests on preprinted stationery, once a day.

ii. Information screens–although the department has produced a handbook, giving details of reference intervals, methods of carrying out certain test procedures, and providing general comment on many tests, such printed material becomes out of date alarmingly rapidly, necessitating sending out amending circulars, etc. We have recently introduced information screens on VDUs, covering much the same ground as the handbook, which the laboratory staff can readily update rather than have to wait for the next printing of the handbook.

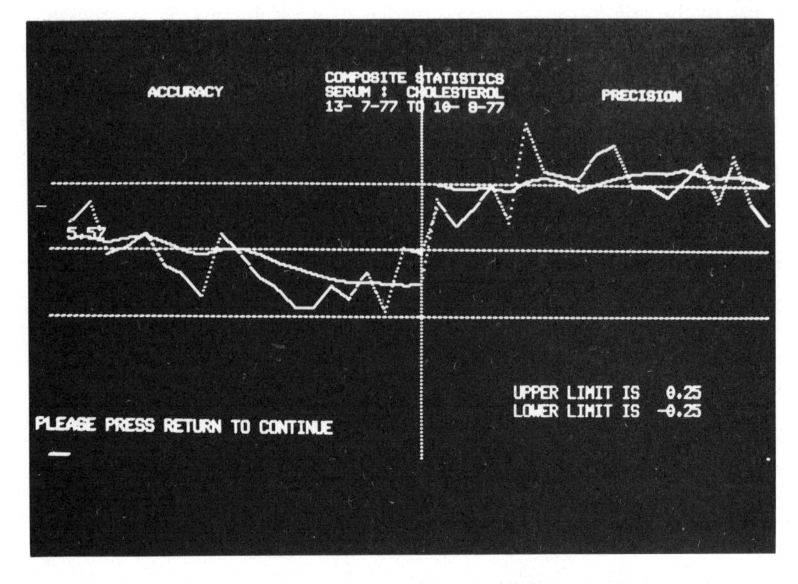

Figure 5. Quality control data on a Digital decscope VDU

EGB DREU [illegible]

CLINICAL CHEMISTRY REPORT - SECTION ONE

COLLECTION		[--------URINE OR OTHER FLUIDS							---------]	[-------SERUM			-------]
DATE	TIME	Type	Vol	Hours	Na	K	Urea	Prot	Cr Cl	Creat	Na	K	Urea
30/05	NG	UR	1760	24	125	62							
30/05	NG									88	136	3.5	5.4
30/05													
30/05													
04/06	NG									93	134	3.4	5.2
04/06													
04/06													
04/06													
04/06													
10/06	NG									86	132	3.3	5.3
A.C.L.+/-	//////////									7	3	0.1	0.3
S.D.D.	//////////										-2.3	-2.2	

To receive section two press 'SEND'

Figure 6. Part of a cumulative report on a Cossor VDU screen

CONCLUSIONS

In our experience VDUs have proved remarkedly trouble free, and usable not only by trained data processing operators but also, by general laboratory staff. The use of VDUs is very simple indeed, though the speed of operation depends on the type and on the complexity of the procedure involved. There have been few complaints about legibility of data on the screen, though the larger Cossor screens are preferred to the smaller decscope units, and green is marginally preferred to grey as a colour for data on screens. We find that VDUs as input devices are excellent, rapid in the hands of trained operators, almost free of mechanical breakdown, with extremely low error rates, though naturally this depends on the skill of the operators. Complementary to tape and punchcard systems, therefore, VDUs have a place as input devices, though this process is dependent on the care of individual operators, and there is a danger of breakdown if the system is 'too simple' and boring.

As output devices, VDUs have an even bigger role to play, widely approved by both laboratory and clinical staff, though it must be appreciated that acceptability depends not so much on the VDU as on the computer system and on its response time. They have a very useful role for presenting data of transient importance, obviating the need for more and more paper which is a frequent criticism of computer systems, and an additional role for presenting data of more permanent value. Here however we find that clinical staff still demand as many printed reports as ever, though our records show a steadily increasing rate of use on VDUs by clinical staff, and possibly, in time, some reductions in printed reports could be contemplated.

ACKNOWLEDGEMENTS

Both the departments' own work on computing, and the hospital real-time computer system have been dependent on the support of the Department of Health and Social Security. Inevitably, laboratory computing depends on team work, and it is a pleasure to acknowledge the debt due to Mrs Margaret Peters, Principal Computer Scientist in our laboratory and her staff; to Mr P M Hills, Director of Computing and Mr S W Sargent, Chief Systems Analyst and many members of the Computer Department of the Queen Elizabeth Hospital; and to Mrs Maeve George and Mrs Linda Stables, successive supervisors, and the members of the Pathology Data Unit.

REFERENCES

1. Whitehead, TP, Becker, JF and Peters, M (1968) In *Computers in the service of medicine* (Ed) G McLachlan and RA Shegog. Oxford University Press. Page 115.
2. Peters, M (1976) *Biomedical Computer Technology 1,* 25
3. Whitby, JL and Blair, JN (1976) In *Automation, Mechanisation and Data Handling in Microbiology* (Ed) A Baillie and RJ Gilbert. Academic Press, London and New York. Page 23
4. Eason, KD (1976) *Computer Journal 19,* 3
5. McSwiney, RR and Woodrow, DA (1969) *Journal of Medical Laboratory Technology 26,* 340
6. Bold, AM (1976) *Lancet 1,* 951

THE BARCODE–A UNIVERSAL SYSTEM OF IDENTIFICATION IN THE MEDICAL LABORATORY

Detlef Laue, MD
Institute of Clinical Chemistry and Nuclear Medicine,
Cologne, Germany

INTRODUCTION

In hardly any branch of medicine are the problems of identification so varied as they are in the medical laboratory.

The most diverse assortment of substances such as blood, plasma, serum, urine, faeces, liquor, sputum and stomach or duodenal juices, with or without additives such as anticoagulants, that have been acquired at different times of the day or collected during various periods, before and after treatment with certain test-substances, reach the laboratory in miscellaneous receptacles such as capillaries, smears, test-tubes, special containers, bottles or wrapped in paper, as are stones.

It is essentially an organisational task of the laboratory to label the specimens clearly as soon as they reach the laboratory with the name of the patient, date of birth, address, age, sex, type of test orders, source of specimen, specific conditions of specimen, name of referring doctor, ward, health insurance plan, etc. During the ensuing manual or mechanical procedures inside the laboratory, the specimens must remain clearly identifiable to ensure the classification of analyses and results to the correct patient. Even in extensively mechanised laboratories, manual work is still required to ensure accurate identification of specimens during all stages of processing. Whenever the specimen must be transferred to a secondary tube, or put into an instrument, or directed through a process of exploratory tests then manual handling is unavoidable, and carelessness could cause confusion. A great part of laboratory work, notably rare, difficult or qualitative methods, is not suitable for mechanisation. In such cases only strictly adhered to procedures can prevent mix-ups. A continuous system of identification for all laboratory areas that work without human assistance does not exist, and never will be able to exist, simply because of the cost.

Nevertheless, it is essential to employ a system of identification in the laboratory. The traditional hand-labelling of containers is objectionable due to errors in reading or writing, the general inconvenience of illegibility and it is also very time consuming.

Most of the available specimen identification systems which use special containers are limited in their implementation. In this manner, both the Eppendorf and the Silab systems restrict themselves to system-specific containers. Other sample containers, such as hematocrit-capillaries, erythrcyte and leucocyte diluting containers, urine bottles and the like, cannot be incorporated into these systems. Here, adhesive labels would be useful. Machine-written labels are plainly superior to hand-written ones.

The aim must be for visual and machine-readable labelling of containers and other records pertinent to the identification process. This is indispensable to efficient automation. The machine code to be aimed at should be able to be used universally in laboratories, be visually legible, be able to be read without error, as well as being functional at all stages of the testing. These requirements are met by the Optical Character Recognition (OCR) code and the barcode.

The OCR code may, in the first instance, appear more attractive, because it is immediately visually legible. At the same time, it should be remembered that the code numbers are meaningless to most users in the laboratory area, and that the code's meaning must be printed out alongside or on a second line. Therefore, as far as visual legibility is concerned, the OCR code offers no particular advantage over the barcode.

Practically every typewriter which could be fitted with the conforming print-types could produce OCR labels. That would be an inexpensive label printer. However, if the labels have to include a line of interpretation, which for reasons of speed must be typed on the same occasion as the OCR code, then a special printer would be needed which would not be much cheaper than a barcode printer. Very complicated, and therefore expensive, is the OCR reading-unit. Moreover, as long as the reading-pistol is manually directed over the code, considerable precision is demanded in the handling of the reading process.

THE BARCODE

The barcode, in relation to the OCR code, presents no significant disadvantages. The numeric content of the code is printed out synchronously in the underlying lines, and is just as restrictively 'readable' as the numeric content of the OCR code.

The real advantages of the barcode over the OCR code are its reading dependability and the modest price of its reading-unit. These two reasons prompted us to install the barcode in our laboratory.

The barcode serves to read numeric information mechanically. Every cipher from 0 to 9, and from A to F is indicated by four bits. Each decimal digit is represented by a binary coded decimal number (BCD) of four bits. The bits are of varying thicknesses, whereby a thin bar corresponds to a binary 0, and a thicker bar to a binary 1. Moreover, there is another combination of the bars for special purposes such as the check code, the termination block, and the reverse start code, which enables a label to be read correctly 'backwards'.

The label structure which we use comprises 13 characters, 9 of which are useful data. The first character, located on the left, contains a start code which electronically initialises the data pens. This is followed by the data block of 9 useful characters, thereupon the check code consisting of 2 CRC test code characters, and a reverse character which acknowledges the finished reading or the return reading.

The operation of the data pen is dependent upon the difference in reflected light from black and white areas in the code pattern. The data pen contains an infrared light emitting diode (LED) and a phototransistor. The white and black alternations of reflected light are registered, and the time between them put in ratio. The information 1 or 0 is derived from the ratio of white time to black time. For that reason, a wide range of reading speeds is tolerated.

The hard fibreglass stylus fitted to the light-pen scratches through contamination and the choice of light wavelength to operate the pen makes it relatively insensitive to contamination-induced noise. In tests with around 500 labels severely soiled with blood or serum, not one reading error was registered.

The readings which are not accepted are indicated by a buzzer. The readers are very inexpensive. The data pen with the LED costs 160 DM, the complete reading unit 2,500 DM.

Because we require many data pens and only a few printers in the laboratory, the barcode is significantly cheaper than the OCR code, the complete reading unit alone of which is approximately three times more expensive.

The barcode labels are produced by special printers that work much like a drum-printers, and have simultaneous alphanumeric characters. Thus, the content of the barcode and additional information can be printed out in clear text. The label printer costs roughly 31,000 DM. In the laboratory only a limited number of printers are required at the reception desk. For example, we use 5 label printers in our laboratory.

In practice, the usage of the barcode works in the following way. We give the referring doctor or ward our request forms which specify the types of analyses in barcode, and a strip of barcode submitter labels which are to be stuck onto the forms where marked (Figure 1). The submitter label shows the code number, beside the name and address of doctor in short form, under which the doctor or ward is kept in our data file. The most important patient data, which is surname, forename, date of birth and address is filled out by hand on the form. The referring doctor can then signify the health insurance plan, so that the bill can be drawn up according to the corresponding tariff.

The tests requested are marked with an 'X'. Additionally, the type of material collected is crossed in and perhaps the 24 hour urine output recorded.

On arrival at our reception desk in the laboratory, the surname, forename and date of birth are fed into the computer at the terminal; and, the submitter, the health insurance plan, the type of substance and the tests requested are read in with the data pen, a procedure which goes extremely quickly.

Immediately thereafter, the required quantity of barcode labels for the analyses are printed out. The printer needs 18 seconds for the production of 10 labels. Each one is an individual label with the ID number assigned to the patient by the laboratory. This label, which contains the surname, forename, sex and ID number and also the ID number in barcode is stuck to the request form. The other labels, one for each container that will be required to perform the tests, contain the ID number and the analysis requested, both in barcode. These are stuck onto the various specimens including the serum tube after centrifuging. The specimens reach the various laboratory departments with these labels.

Figure 1 (Opposite) Part of a test request form showing the barcodes preprinted on the form and also, in the top right hand corner, a submitter and a patient label.

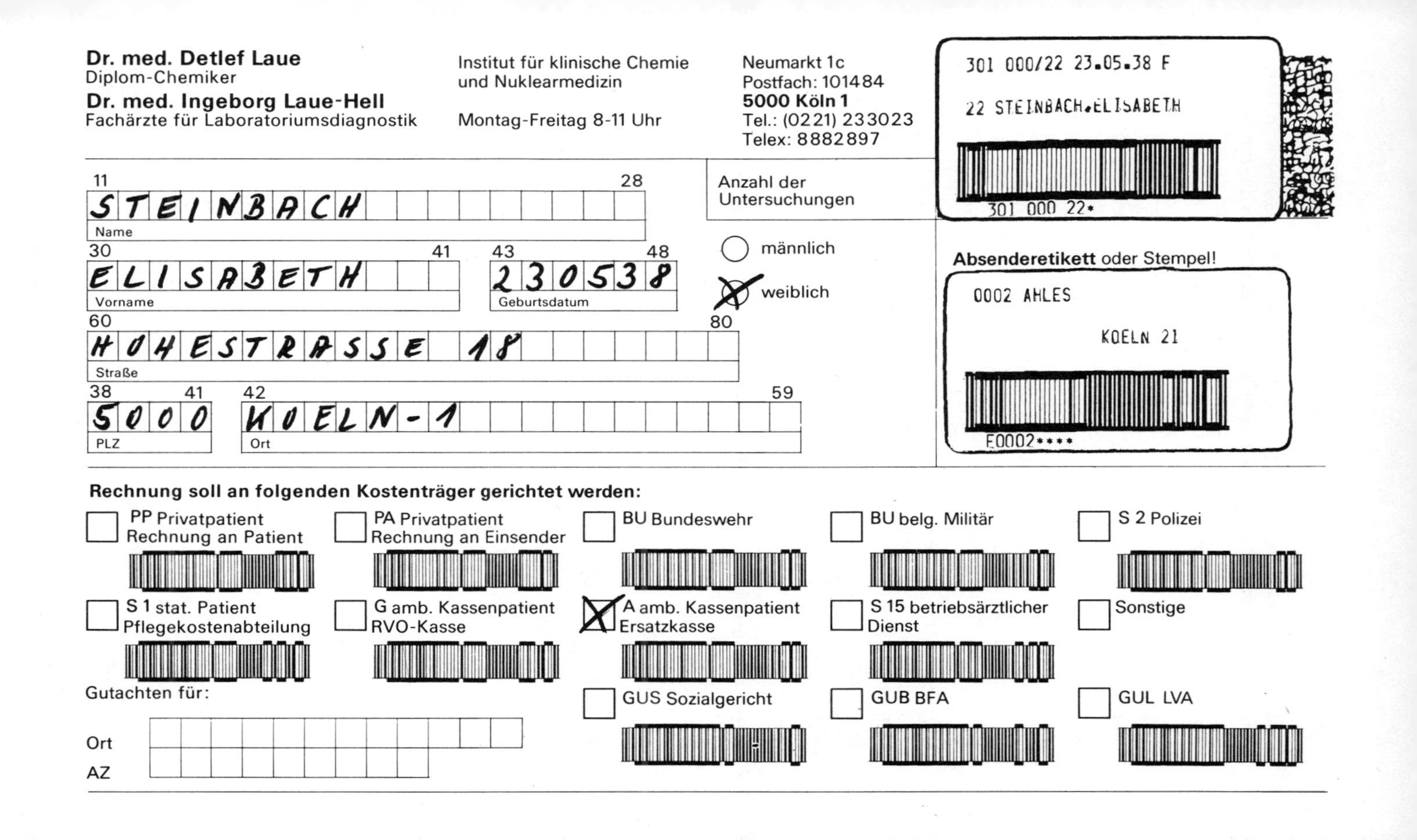

Dr. med. Detlef Laue
Diplom-Chemiker
Dr. med. Ingeborg Laue-Hell
Fachärzte für Laboratoriumsdiagnostik

Institut für klinische Chemie
und Nuklearmedizin

Montag-Freitag 8-11 Uhr

Neumarkt 1c
Postfach: 101484
5000 Köln 1
Tel.: (0221) 233023
Telex: 8882897

301 000/22 23.05.38 F
22 STEINBACH.ELISABETH
301 000 22*

11 STEINBACH 28
Name

30 ELISABETH 41
Vorname

43 230538 48
Geburtsdatum

Anzahl der Untersuchungen

○ männlich
☒ weiblich

60 HOHESTRASSE 18 80
Straße

38 5000 41
PLZ

42 KOELN-1 59
Ort

Absenderetikett oder Stempel!

0002 AHLES
KOELN 21
E0002****

Rechnung soll an folgenden Kostenträger gerichtet werden:

☐ PP Privatpatient Rechnung an Patient
☐ PA Privatpatient Rechnung an Einsender
☐ BU Bundeswehr
☐ BU belg. Militär
☐ S 2 Polizei
☐ S 1 stat. Patient Pflegekostenabteilung
☐ G amb. Kassenpatient RVO-Kasse
☒ A amb. Kassenpatient Ersatzkasse
☐ S 15 betriebsärztlicher Dienst
☐ Sonstige

Gutachten für:
☐ GUS Sozialgericht
☐ GUB BFA
☐ GUL LVA

Ort

AZ

We no longer set up work lists. In this way the laboratory is independent of the computer and the technicians can start with the tests at any time, and it is not necessary to follow any set sequence of samples. As soon as the samples are ready the barcode on the test-tubes can be read in. Thus, the computer acquires information about the type of investigation, the sequence of the samples and the position of the controls and standards within the series. The automation is set in motion by a start signal, and summons the requested program by the reading in of the corresponding barcode command. The initial calculation takes place in a microprocessor which prints out the calculated results via a terminal. At the same time, the results are stored away on a floppy disc.

A limitation of the barcode, and the OCR code as well, is the fact that the labels are too large to be stuck onto microlitre containers, and at present we do not have mechanical instruments that allow automatic scanning of the test-tubes during the measuring process. But we believe that the universal application possibilities of the barcode system outweigh the mentioned drawbacks. In particular, we can identify every laboratory slip or paper or document with the machine-readable barcode labels. (See also Labarrere et al in this book.)

Because the barcode labels are too large for the microlitre containers, the upper portion is perforated and can be ripped off and stuck onto the microlitre containers. Thus, the system furnishes identification labels, which of course are only visually legible, for small containers as well.

Long term tests of barcode systems in libraries, supermarkets and warehouses, show that there are still concealed problems.

DATA PENS

We use the Plessey data pens. They have a fibreglass style which is held in a thin steel tube. The mechanical abrasion of fibreglass is faster than the abrasion of the encasing tube so that, when the initially somewhat protruding fibreglass point is worn, a concave hollow forms on the convex tip. In this cavity dust particles could accumulate which would hinder the passage of infrared rays and lead to reading difficulties. However, it is possible to clean the hollow, or sharpen away the metal with the finest grade of emery-paper. With us, the durability of such a pen is roughly 12 months. The infrared reading-unit of the pen is as reliable as other modern electronic devices.

LABEL PRINTERS

As we reported years ago, we originally began work with Anker label-printers; but, we were unable to extend our system because Anker went bankrupt, and the production was closed down. That was a bitter and expensive experience for us. We then changed over to the label-printer distributed by Plessey, and produced by Intermec, Washington, USA.

Whereas, the first printer delivered by Plessey worked for a year without any mechanical or electronic failures, another one of three delivered this year, apparently from a new production series, has caused considerable problems. The main problem with this new series seems to be an excessive development of heat in the mechanical area of the printer (where the motors and transformers are), and an inadequate ventilation of the chassis. As a result, misalignment of the printing mechanism followed and corrosion of the ball bearings which support the print drum set in.

Since there were at the time only six Plessey label-printers in Germany, Plessey maintained no part replacement centre. And because they were unable to offer any efficient maintenance service in Germany until the summer of 1977, we built ventilators into the printers in our own workshop, and replaced the ball bearings with new ones which we oil regularly.

Once a print-hammer broke and the replacement took 8 days to deliver from USA. Subsequently, it was very difficult and time-consuming to realign the print-hammers.

On the electronics side, the printers presented no extraordinary problems. Small failures, such as defective modules, which happen with all electronic machines, could be handled by our own workshop.

A serious problem, as yet unsolved by the manufacturer, is that the printing process is associated with an unbearably high frequency noice, which cannot be tolerated in the patient receiving areas or in the laboratory. For that reason, we have housed the printers in large movable crates that are sound-proofed and ventilated.

While considering the disadvantages of the printer, we should not forget the high price of 31,000 DM. If compared with the price of an efficient fast-printer today, it should be possible for the label-printers to be offered substantially cheaper.

INTERFACE

As interface between our modified Dietz Computer 523 and the terminal with console typewriter, data pen and label-printer, we use a microcomputer system, developed and built in our laboratory. It uses the MC 6800 from Motorola as microprocessor. The cycle time in this system is 2 microseconds, giving a command execution time of between 4 and 24 microseconds. Our microcomputer has RAM's of 16k byte and PROM's of 32k byte, and is still capable of expansion.

We now have the following instruments connected to the microcomputers in our laboratory:

2 displays
12 teletypes
2 modems
Eppendorf photometer
2 line-printers from Tally
1 high-speed-printer from Dataproducts with 600 lines a minute
1 punch-tape reader, and
1 computer-to-computer connection.

To some extent, floppy discs are built into the microcomputers for initial processing, and for holding microcomputer programmes, in as much as this is required in the laboratory area. All data input terminals, and that includes displays and teletypes, are equipped with Plessey data pens, as are the measuring monitors in the laboratory. Finally, the 5 label-printers are connected via microprocessors.

At present we have 10 microprocessors connected to our central computer which, with 12 process levels and the vast autonomy of the microprocessors, permits a smooth flow of data. It is thus possible by computer control to operate the label-printers at the same time, with very little waiting time. We do not use the label-printer keyboard. The operation of the label-printer and the data input is performed by teletypes via the computing system.

CONCLUSIONS

In summing up, we can say that a universally applicable system of identification in the laboratory can be established with the barcode. Such a system allows not only patient identification of specimens, but also provides, as we have shown, data input of patient ID number, submitter, health insurance plan, type of substance and type of analysis requested—all according to the marking of the request forms, laboratory records, specimens, etc.

Apart from relatively few short-comings, the hardware proved its worth. The data-pens and the label-printer were integrated into our computer system by our own efforts.

The installation of the barcode has significantly expedited the data going into the computer, and seems to have considerably reduced the rate of errors.

OPTICAL MARK READERS IN CLINICAL MICROBIOLOGY

J.C. Coleman and A.W. Linington
Departments of Medical Microbiology and Medical Computing Science, Charing Cross Hospital, London

INTRODUCTION

In March 1968 the Charing Cross group of hospitals submitted a proposal to the Department of Health and Social Security seeking authority and financial support to form a Department of Medical Computing Science. The proposal was accepted in May 1968.

A pilot scheme for computer assisted microbiology reporting was introduced in August 1972 (1). This system used one Termiprinter and an off-line optical mark reader (OMR) linked with the Leasco time-sharing computer bureau. In August 1973 the system was transferred to the Comshare bureau which gave experience of the Rank Xerox computer, in anticipation of the commissioning of the live in-house computer at Charing Cross Hospital in February 1974.

The Microbiology laboratory has been using OMR worksheets since 1972 to record test results and enter them into the computer. The computer assisted microbiology reporting system at this hospital processes some 90,000 requests each year.

THE PRESENT SYSTEM

The system used in Microbiology (Bacteriology and Virology) is outlined as follows:

1. Punch request details on paper tape, or keyed in on-line
2. Read paper tape into computer
3. Laboratory mark results of tests on worksheets
4. Worksheets read into computer using on-line optical mark reader
5. Virology reports edited on the Visual Display Unit
6. Reports printed by on-line laboratory terminal
7. Reports checked, signed and returned to requester

On arrival at the laboratory, specimens and request forms are checked by laboratory staff to ensure that patient, specimen and request information is correct.

At this point the request is allocated an OMR worksheet which bears a serial number by which the request will be identified both within the laboratory and in the computer. The departmental clerical staff input the request data on a teletype machine either by generating a paper tape or directly on-line. The request data comprises the following information: worksheet number, patient hospital number, patients name and location, a coded description of the nature of the specimen and a consultant identification code.

The laboratory technicians receive the specimen, request form and worksheet. The worksheet number is the number by which the specimen and its subsequent cultures and subcultures are identified in the laboratory. The results of the tests or cultures are recorded by the technicians making bar-marks in the appropriate cells in the worksheet. The organism codes are based on those of the Association of Clinical Pathologists Working Party (2). The OMR form currently in use provides space for four different organisms to be coded, quantified and sensitivities recorded.

The completed worksheets are read into the Rank-Xerox Sigma 6 computer by a Data Recognition RT 8301 OMR on line via a 1,800 baud line. An OMR is a device of photoelectric cells which detects the presence of bar-marks in specific areas of a predesigned document. The signals so generated are transmitted to the computer and interpreted by a specific program which integrates the information from the worksheet with the patient identification which may have been entered some days, or even weeks, earlier. The worksheet number is used as the key for this purpose. At this stage it is possible to call up virology reports onto a visual display unit (VDU) screen for editing and correction.

The reports are printed by an on-line ICL Termiprinter in the laboratory, checked, signed and returned to the requester. Additional computer output e.g. alphabetic day-lists of requests, alphabetic lists of a day's results, weekly summaries, cross-infection lists, are printed by an on-line printer in the Department of Medical Computing Science.

ALTERNATIVES TO OMR INPUT

A method of input of laboratory data into a computer requires to be rapid, to involve the minimum of human effort, to be accurate, and flexible. The optimal system is an automated testing machine which interfaces directly with the computer. Despite many attempts at automation the basic steps in diagnostic medical microbiology have been unchanged for decades. Manual testing of specimens produces cultures which are inspected and identified by technicians. The vast majority of results are non-numerical and generated manually. Furthermore, the interval between receipt of a specimen and the production of a final report may be measured in days or even weeks.

The results may be handwritten in full and then transcribed for the computer by a secretary or even by the technician. This increases the departmental workload and does not alleviate the transcription errors inherent in a conventional system. The technician may enter results directly into the computer via a teletype or VDU. This is a slow process, requiring a number of terminals, and presupposes a degree of keyboard expertise which is not normally associated with the training of laboratory technicians. Input by punch card is likewise subject to certain limitations. Technicians may produce source documents from which punch cards are produced by a trained operator, a system which retains the possibility of transcription error, or technicians may themselves produce punch cards manually with coded results, a time-consuming process. There is little experience, in microbiology, of optical character readers as a method of input.

ADVANTAGES OF OMR INPUT

The principal benefits of optical mark reading are that it reduces the number of transciption errors at the report stage and that input is rapid.

There are many factors which contribute to the success of an OMR system. These may be divided into factors involving software and those of hardware.

The design of the OMR worksheet is of paramount importance. It is essential that the worksheet be acceptable to the technicians who are required to complete it. A form which is poorly laid out, or is very complex will be either rejected out of hand or be a significant source of coding errors. In its final form a worksheet should be so designed that groups of tests which are performed together, or sequentially are arranged in such a manner. The design of a worksheet should involve close co-operation between laboratory staff, systems analysts and systems designers. The result of this collaboration should aim to produce a worksheet which is not only highly acceptable to the laboratory staff but which may also act as an *aide-memoire* for the technicians, thus ensuring that all tests and investigations on a specimen are performed and duly recorded. Within this design envelope it is essential to remember that the minimum of extraneous effort is required from the laboratory staff. In this context it must be appreciated that the computer program can be so designed that a single bar mark can code for up to one or even several sentences of a report. The drop-out printing which is not read by the machine should be designed so that in the event of computer failure the bar marked codes can be readily interpreted by laboratory and clinical staff, permitting a manual back-up facility.

A study by a Department of Health and Social Security Computer Evaluation Unit at Charing Cross Hospital revealed that 13 out of 14 technicians agreed that bar marking was straightforward. All 14 technicians stated that bar marking was quicker than making abbreviated notes.

The worksheet in use at Charing Cross is comparatively simple when compared with those in use at other centres. It is entirely contained on a single sheet of A4 sized paper.

Rapidity of input is a major feature of an OMR based system. Automatic supply and feed mechanisms can feed documents at the rate of 300-2,500 sheets per hour depending largely upon the size of the document. A random examination of our system at Charing Cross has revealed that documents are processed at rates varying between 10 and 18 per minute. The variation is almost entirely due to variations in the response time of the central processing unit (CPU), and the dexterity of the operator.

Using more complex worksheets the rate of input may be slowed marginally but it is possible to devise more ambitious programs. If the results of individual tests e.g. Gram stains, sugar fermentation reactions etc. are included on the worksheet it is possible to write programs for computerised identification of microorganisms. Such programs are already in use in some laboratories. From the same input it is possible to monitor accurately the use of individual culture media and test reagents. This information can be used advantageously in prospective bulk purchasing of media and reagents. This system enables the staff of the laboratory, additionally, to assess accurately the cost of tests performed in the laboratory over a period of time. This is of value when attempting to produce realistic departmental budgets.

A final advantage of OMR input becomes apparent during failure of the CPU. The output of the machine may be used off-line to produce a punched-paper tape. When the system is eventually restored, the paper tape may then be read into the CPU as a batch job.

DISADVANTAGES OF OMR INPUT

The high cost of the mark reader itself is the major disadvantage of a system based on OMR input of laboratory results. The prices of units range from £5,000-£65,000. The possibility of machine breakdowns, which in our experience are infrequent, does impose the necessity of having a second, back-up, machine readily available.

The cost of worksheets, which must be produced by specialised printers is also an important revenue consideration in a busy laboratory. The Charing Cross worksheet costs approximately 3.5 pence.

Critics of OMR input systems point to the inflexibility of the system. It is true that a preprinted worksheet does impose some constraints upon the user. However, careful design of the worksheet goes a long way towards mitigating these effects. Some laboratories achieve a high degree of flexibility, particularly when reporting antibiotic sensitivities, by the use of suitable templates overlying blank cells on the worksheets. The inflexibility of the system does become particularly exasperating when the introduction of new tests, or the recognition of new pathogens, demands that the worksheet be drastically redesigned before the new tests etc. can be reported by the computer. The provision of spare cells or the reallocation of redundant cells on the worksheet may overcome this problem.

It is highly desirable that those whose duties require them to fill in the worksheets should have some training. We have found that laboratory technicians have no difficulty with the worksheets. They are aware that it is possible to make errors when bar marking. Out of 14 technicians, 9 felt that it was easy to make a careless mistake, and that misalignment of individual bar marks could also be a source of error. On the other hand, an attempt to produce and use an optically mark readable request form was a failure. It proved impossible to provide training or supervision of the medical or nursing staff who originate the requests for the laboratory. Only 18 per cent of the request forms received were acceptable to the machine by reason of bar marking errors alone. Many were rendered unacceptable by virtue of the traumata which befell them during their passage from ward to laboratory. The worksheets which are completed by the laboratory staff are associated with an obvious and readily correctable rejection rate of 3-4 per cent. The commonest errors are due either to the use of a worn out felt-tip pen or to encroachment across adjacent squares with one mark.

It is highly desirable that a program which uses a purely numeric code for the description of microorganisms should include a check digit in these codes. Without such a precaution it is possible to make coding errors when completing the forms. When interviewed, the technical staff were all aware of this problem, which in itself may reduce the likelihood of such accidents.

CONCLUSION

After five years experience of OMR input of laboratory data we feel that in a busy laboratory the advantages of the system outweigh the disadvantages. The worksheets are acceptable to technicians who as responsible trained personnel, are aware of the possibility of mismarking. The clerical staff find the system easy to operate. Those constraints imposed upon the system by the use of a preprinted worksheet produce a report form of standardised format which is very acceptable to the clinical staff.

REFERENCES

1. Andrews, HJ and Vickers, M (1974) *J. clin. Path., 27*, 185
2. Association of Clinical Pathologists (1968) *J. clin. Path., 21*, 231

OCR READERS

J Dudeck and H Michel
Institut für Medizinische Statistik und Dokumentation
der Justus-Liebig-Universität Giessen

INTRODUCTION

Problems of data input are of high importance in different fields of medical data processing. The evaluation of costs and quality of medical care depend to a high degree on easy and low cost data acquisition procedures. Laborious procedures for data input are often the limiting factor to the user acceptance of computer systems in medicine.

Optical mark and particularly barcode readers opened new dimensions of data acquisition. From our experience with barcode readers I nearly completely agree with the evaluation presented by Dr Laue. The main disadvantage of barcoding is that the machine readable part cannot be easily deciphered by eye and therefore a duplication in normal alphabetic characters is necessary.

OCR-A numeric	0123456789⑀⑁⑂
OCR-B numeric	1234567890
Farrington 7 B	0123456789
Farrington 12F	0123456789EP-H
407	0123456789
E13B	0123456789⑆⑇⑈⑉
1428 numeric	0123456789 ./-+
OCR-A alpha	ABCDEFGHIJKLMNOPQRSTUVWXYZ {}%?&£"*:;=
hand print	0 1 2 3 4 5 6 7 8 9 C S T X Z

Figure 1. Optically readable character sets provided by the Keytronic reader

For more than twenty years attempts were made to recognise characters by technical devices. To diminish the pattern recognition problems and to improve the reliability of the devices, special character sets were developed which improved the discrimination of different characters. The best known are the OCR-A letters. After a short time of adaptation these characters can be easily read by eye. Several other character sets were developed which are already used in industrial environments. More like normally used letters is the OCR-B character set, which also provides numeric and alphabetic characters (see Figure 1).

The development of integrated electronic circuits has diminished the cost and dimensions of the pattern recognition equipment. For nearly three years, mobile, handy and inexpensive optical character readers have been available. One of the first readers was the 'reading pistol', the wand reader produced by Recognition Equipment. It is now available for OCR-A and OCR-B numeric characters (Figure 2). A more recent development is the data wand reader provided by Keytronic which is able to read the different character sets shown in Figure 1, including OCR-A alphabetic and hand-written characters. It is equipped with trading rollers and with up to five push button keys which can provide different functions such as carriage return, line feed, tab, etc.

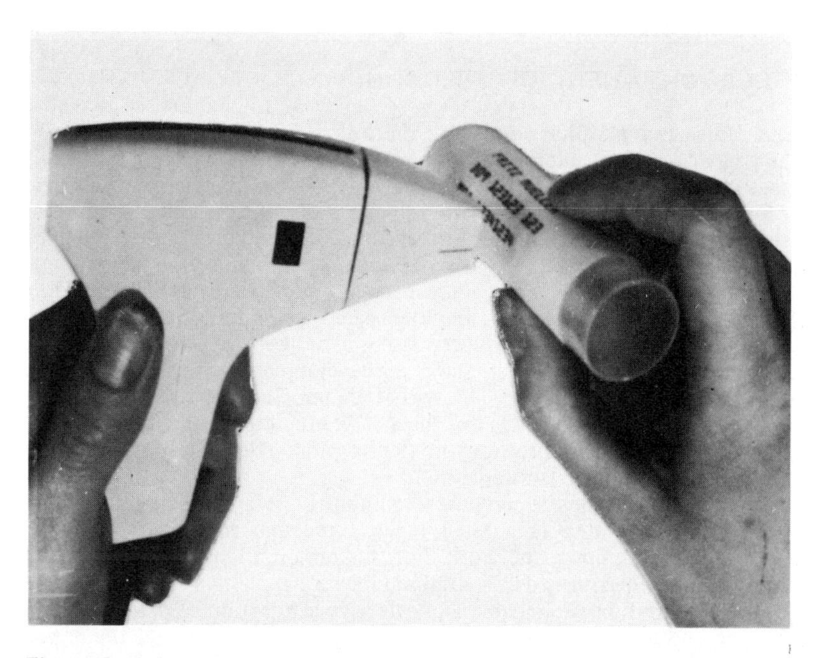

Figure 2. OCR wand reader. Labels on either forms or sample tubes can be read reliably

The main advantage of the character reader over barcoding is the higher information density on the information carrier. The comparison of the information density of optical marks, and barcodes in information density per cm^2 is as follows:

optical marks	1-2 bit/cm^2
barcode	10-15 bit/cm^2
OCR numeric	15-30 bit/cm^2
alphanumeric	25-30 bit/cm^2

This does not include the space necessary for the additional printing of human readable characters on optical mark or barcode labels or forms. A high information density makes possible the application of small labels for the transmission and acquisition of data.

The performance of optical character readers is very reliable. They detect the characters either correctly or reject the reading completely. In this case the reading procedure has to be repeated. During several thousand carefully analysed readings using clean and spoiled labels we only detected one case of a misread character caused by slightly blurred characters.

As examples of the application of optical character readers in the medical environment we would like to describe two projects in Germany.

The first example describes the patient identification and data acquisition system which is going to be introduced in our University Hospital in Giessen.

The second example is a pilot study on the applicability of plastic cards with OCR-B characters in the public health service carried out by Mannesmann Corporation in cooperation with the Federal Ministry of Arbeit und Sozialordnung.

OCR FOR PATIENT IDENTIFICATION AND DATA ACQUISITION

Our University Hospital consists of more than 20 clinics distributed over an area of approximately one square mile. The hospital has about 1700 beds and several large outpatient clinics. For nearly ten years a communication system for laboratory data, using mark-sensing-cards has been operational (1). The performance of this system is very reliable, but it no longer meets the goals of modern communication, since there is a time delay of several hours between processing and distribution of data. For this reason a computer based communication and information system is being introduced using a network of two Modcomp II computers and, in the first stage of development, one Modcomp IV computer as host computer. The two Modcomp II computers are located at the clinical laboratory. One of them serves the inpatient and outpatient admission at different locations in the hospital. The other one processes the data of the clinical laboratory.

The computers in the network communicate via the Maxnet IV network operating system (Figure 3). It was impossible to connect all wards on line to the computer network. For this reason a large amount of data input has to be performed at centralised locations.

In this situation the reliable and effective acquisition of patient identification and data is of great importance to the performance of the entire system. Every day, several ten thousand times, the computer has to identify data and to allocate them to patients. The correct input of the patient identification requires a certain amount of time which is always underestimated. To shorten this time and to improve the reliability of acquisition we investigated the performance of some optical reading devices. More than three years ago we carried out one of the first systematic studies on the application of barcodes (2). The results were very promising. The reliability of the barcode reader was surprisingly high. We could not detect any erroneous reading.

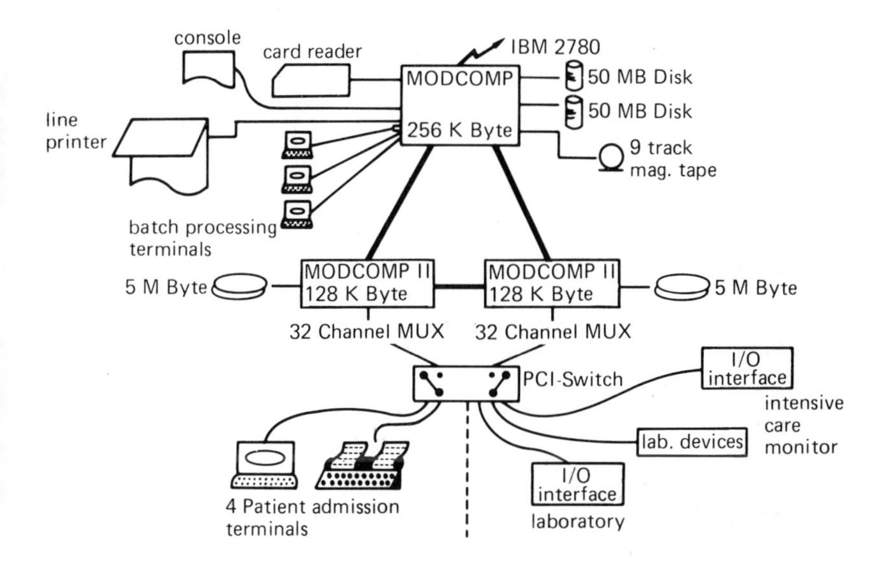

Figure 3. Computer network of the Giessen University Hospital. The two Modcomp II are located in the clinical laboratory. During clinical routine one computer serves the clinical laboratory, the other one the patient admission terminals. They are connected using Maxnet IV network operating system. In the case of breakdown of one computer the other one can take over the most important tasks of both systems.

Just a few days before we had to come up with the final decision about the equipment to be introduced, we had the possibility of testing the OCR-A wand reader. We were very sceptical about the expected performance and were very surprised on discovering the same reliability. The reader could not be disturbed by soiling the characters with blood and other specimens used in the laboratory. Therefore we decided to go into this new field and to introduce the OCR in our systems, particularly because of the higher information density.

On first admission to our hospital each patient receives a unique personal number which is retained by this patient at all times. This patient identification number has ten digits. It consists of the birthday (six numbers in the order day, month, year) two numbers as code of the name (maiden name), one additional number and a self check number (modulo 11). The additional number separates two or more patients with the same birthday and name (e.g. twins). It also marks the sex of the patient. The numbers 0-4 are used for male, 5-9 for female patients. The identification number is easily remembered by the patient as he always knows his birthday and has to remember only four additional digits (3).

```
Y12 054667              W
                     3414
ARNOLD ARIANA
                       15
N2407440053
```

Figure 4. Patient identification label used in the hospital information system at Giessen University Hospital. The top and the bottom line are readable by the wand reader. The leading alphabetical character defines the format of the following numerical characters. In the top line the hospital admission number, in the bottom line the patient identification number is printed. Above and below the name, the number and the telephone number of the ward are printed for conventional communication

On admission of the patient a sufficient number of labels are printed and are sent to the wards where they are stored in the nurses' room and at the patients bedside (Figure 4). The labels are produced using 9 x 9 matrix printers. On the labels the name, sex, ward number and telephone number of the ward are printed to allow conventional communication.

For computerised communication two numbers are used. The patient identification number described above is used for patient orientated purposes. This number does not contain any information about which hospital the patient was admitted to. Since the hospital administration needs this information a second optically readable number was introduced, the hospital orientated admission number. This number is also uniquely allocated to the patient, but it is changed when the patient is readmitted to the same hospital or is moved to some other clinic within the University Hospital.

The labels can be used for the identification of all kinds of requisition sheets and forms and also for the identification of sample and specimen tubes. They support both conventional and computerised communication. By using labels no additional equipment is necessary on the large number of wards in the periphery of the hospital information system.

Having introduced the optical reader for patient identification it seemed reasonable to utilise the same technique for the acquisition of other data, too, especially for data that is generally acquired with the patient identification (laboratory requisition, administrative or billing data). The use of optically readable code should be so arranged that excessive time is not spent looking for the code on a form. The search and read time should together be shorter than the time for entry via a keyboard.

Investigations of the reading times for barcodes have shown that the time depends remarkably on the number of selectable codes on the form (4). Using sheets with only ten selectable codes, the average search and read time per code was 1.2 s. It increased by about 0.5 s per ten additional selectable codes on the same form. Having more than 40 selectable codes the time per code read in increases to nearly 3 s, so that advantages of the method decrease considerably.

There are different ways of providing the codes. One possibility is to print the codes on forms, e.g. requisition sheets (Figure 5). The required tests are marked and at the data acquisition terminal the appropriate codes are read after reading the patient identification code.

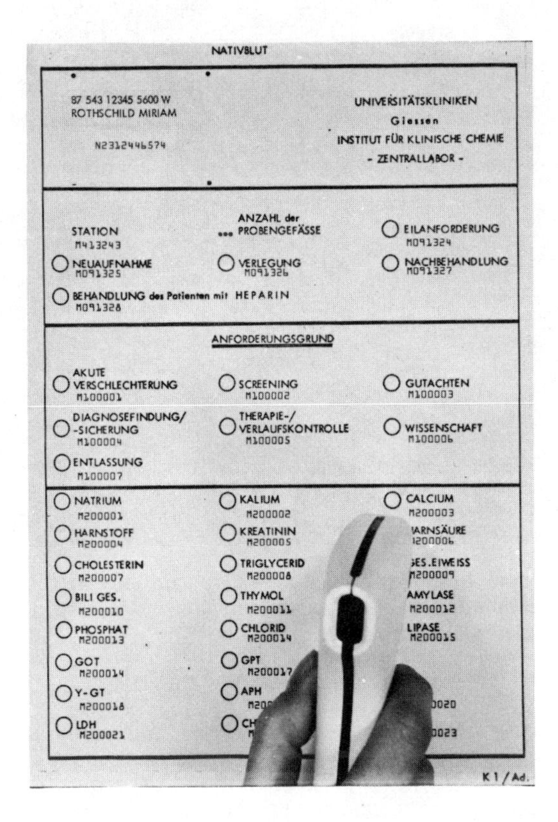

NATIVBLUT

87 543 12345 5600 W
ROTHSCHILD MIRIAM
N2312446574

UNIVERSITÄTSKLINIKEN
Giessen
INSTITUT FÜR KLINISCHE CHEMIE
- ZENTRALLABOR -

STATION M413243
ANZAHL der PROBENGEFÄSSE
EILANFORDERUNG M091324
NEUAUFNAHME M091325
VERLEGUNG M091326
NACHBEHANDLUNG M091327
BEHANDLUNG des Patienten mit HEPARIN M091328

ANFORDERUNGSGRUND

AKUTE VERSCHLECHTERUNG M100001
SCREENING M100002
GUTACHTEN M100003
DIAGNOSEFINDUNG/ -SICHERUNG M100004
THERAPIE-/ VERLAUFSKONTROLLE M100005
WISSENSCHAFT M100006
ENTLASSUNG M100007

NATRIUM M200001
KALIUM M200002
CALCIUM M200003
HARNSTOFF M200004
KREATININ M200005
ARNSÄURE 200006
CHOLESTERIN M200007
TRIGLYCERID M200008
ES.EIWEISS M200009
BILI GES. M200010
THYMOL M200011
AMYLASE M200012
PHOSPHAT M200013
CHLORID M200014
LIPASE M200015
GOT M200014
GPT M200017
Y-GT M200018
APH
LDH M200021

K 1 / Ad.

Figure 5. Requisition sheet used in the clinical laboratory. The nurse marks the required tests on the ward and allocates the sheet to a patient by sticking the identification label on it.

Another means of fast access to the appropriate codes can be achieved by placing sheets with codes beside the terminal with the OCR reader. The operator then reads the patient identification and the data code from different sheets. This type of system is preferred when the selectable code has to be repeated several times on the same form, so that there is not enough space to print it on one page.

In our experience the application of directly readable codes opens a large number of organisational possibilities, facilitating data acquisition in hospital information systems.

The unambiguous allocation of samples and test results to a patient throughout the entire clinical laboratory is today a self-evident requirement of every patient identification system. In our system this problem is solved by using the automatic sample distributor produced by Eppendorf.

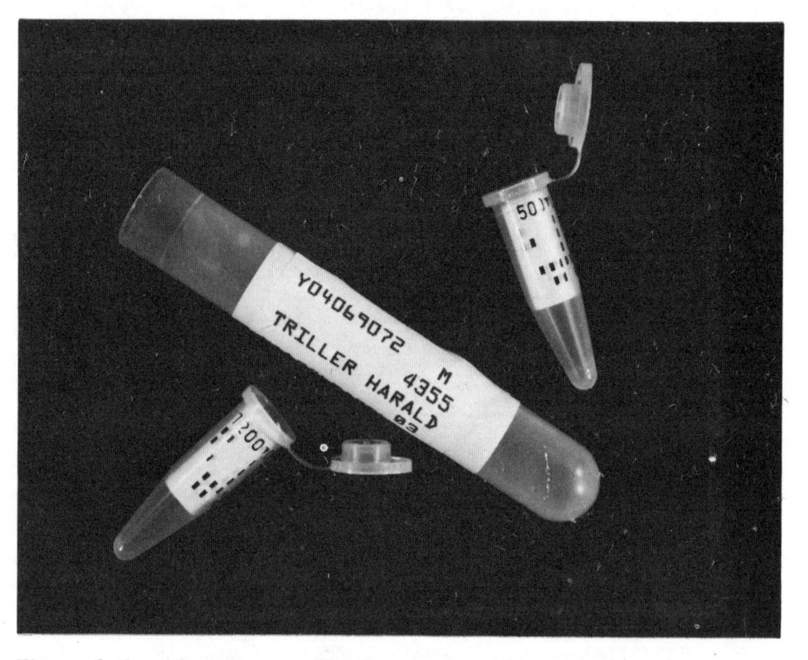

Figure 6. Sample tubes used in the ward and Eppendorf specimen tubes which are automatically produced in the sample distributor by reading the patient identification on the sample tube

The laboratory requisition sheet is marked by the nurse and sent to the laboratory. Using the wand reader, the patient identification, ward number and test requisitions are transferred to the requisition file of the laboratory computer. The sample tube arrives at the laboratory together with or independently of the requisition sheet. After centrifugation the sample tube is brought to the Eppendorf sample distributor, which is connected on-line to the laboratory computer. Using the wand reader the patient identification on the sample tube is also read. In the requisition file the tests required for this sample are found and the devices to which the serum has to be distributed are defined. Controlled by the laboratory computer, one small tube with Eppendorf identification is produced in the sample distributor for each analytic device.

The serum is then manually distributed to these tubes. Since the only tubes in the distributor are those which belong to a specific sample, there is no chance of mixing up samples from different patients.

The Eppendorf tubes are identified by coded numbers (two-of-five code), which are readable by Eppendorf readers at all laboratory devices. The identification number on the Eppendorf tübe consists of the number of the device, a laboratory number and certain additional digits. The laboratory number is unambiguously allocated to the requisition. It was introduced instead of the patient identification since there could be more than one requisition for one patient on the same day.

In this way the unambiguous linkage of the sample to the patient is guaranteed, from the bed of the patient throughout the entire chain of analytical and data processing to the final report.

OCR OF PATIENT PASS CARDS

Another example of the application of optical character readers in the health service is by the German Ministry of Arbeit und Solzialordnung (Labour and Social Relations) In cooperation with the Mannesmann Datenverarbeitung GmbH, Ratingen, W-Germany (5).

During recent years there has been a remarkable increase in the cost of medical care in Germany. This increase had to be arrested otherwise in four or five years people would have to work only for the cost of medical and social care. The problem is also evident in other highly developed countries.

For the evaluation of cost effectiveness of medical care a case orientated analysis of the cost structure of medical care seems to be necessary. Such an analysis requires the recording of prescribed drugs, examinations performed and so on about each patient. The recording of this information by card punching or through the keyboard would require far too much labour. The use of patient pass cards, comparable in physical form to credit cards, is being tested in one region in northern Germany.

Each member of the social security receives a patient card on which the name and address of the patient and insurance number is impressed. On an additional line these numbers are printed for direct reading by optical character readers. To reduce the difficulties, particularly with inexperienced people, the OCR-B numeric characters were used.

Since the numeric characters printed from the plastic cards have also to be readable by optical readers, higher quality standards for impress devices and imprinters had to be fulfilled. The available imprinters for credit cards did not meet these standards. New devices were therefore developed so that the printed characters could be read by optical character readers with very small error rates. In addition to power driven printers, small hand sets, to be used at various locations where only relatively few pass cards are to be produced, are being developed. The power driven devices are very reliable but a few problems are still to be solved with the hand sets.

It is hoped to have the final results of the pilot study by the middle of next year.

Early experience has shown that plastic cards can be used for printing optically readable code can be printed on in exactly the same position on possibilities for the introduction of optically readable codes. One of the main advantages of the plastic cards is that by using imprinters the optically readable code can be printed on exactly the same position of each form. The exact positioning of the code enables the application of fixed-head optical character readers. For this reason it might be possible to combine, for example, the advantages of optical mark and optical character readers in one device, which is not possible with labels.

At present the optical character reader seems to have the most advantages of developments in the field of optical readers. This does not mean that we should use optical character readers only.

The performance of all optically readable code is very reliable. Before introducing an optically readable code into the hospital environment a careful analysis of the pros and cons of different codes is necessary. Up to now no rules have been available to help decision making between barcode and OCR-A and OCR-B readers. The organisational structure of the hospital information system has to be analysed. In particular the following problems have to be taken into consideration:

1. costs and numbers of printers necessary,
2. costs and number of readers,

Table 1 Differences between Barcode and OCR Readers

	Printer	*Reader*	*Information density*	*Alphabetic letters*	*Application of plastic cards*
barcode	special devices necessary Should provide alphanumeric character and barcode DM 20.000	low price DM 1-3000	10-15 bit/cm^2	available	not tested
OCR-A	all 9*9 matrix printers can be used DM 8-15.000	more expensive DM 6-8000	numeric 15-30 bit/cm^2 alphanumeric 25-50 bit/cm^2	available	not tested
OCR-B	using matrix printers a 24*7 matrix is necessary cheap non-matrix printers are supposed to be coming onto the market	similar to OCR-A	numeric 15-30 bit/cm^2	not available	applicable

3. identification carrier (label, plastic card),
4. information density needed on the identification carrier.

Table 1 gives some useful hints on the decision between barcode and OCR readers.

Figure 7. A patient pass with directly readable OCR-B characters in the top line and imprinted OCR-B characters in the second line

CONCLUSIONS

Low cost, easy to use and mobile optical character readers are now available. The performance of these devices is reasonably reliable. They can be used for the solution of data acquisition problems in clinicial information systems and also in the public health service. Technology developments provide us with a large number of different suitable devices. Before introduction a careful analysis of the entire system is necessary so that the best alternative can be chosen.

REFERENCES

1. Hass, H and Kielas, G (1969) *Ärztl. Lab. 15,* 158
2. Dudeck, J and Michel, H (1976) In *Klinisch-stat. Forsch.* (Ed) S Koller and J Berger, Stuttgart, Page 387
3. Dudeck, J and Michel, H (1977) *Proc. Medcomp 77,* Online Conf. Ltd., Uxbridge. Page 65
4. Dudeck, J, Lang, H and Michel, H (1976) In *Interaktive Datenverarbeitung in der Medizin,* (Ed) G Wager and CO Köhler. Stuttgart, Page 295
5. Mannesmann Datenverarbeitung GmbH Entwicklung eines Datenträgers für Versicherten- und Leistungsdaten in der gesetzlichen Krankenversicherung (DAKRAV), Ratingen 1976

Part 5

INTERFACING AND REMOTE PROCESSING

ELECTRICAL INTERFERENCE IN PATHOLOGY LABORATORIES

A.K. Dobbie BSc(Eng), CEng, MIEE
(formerly Electrical Safety Engineer, DHSS)

INTRODUCTION

The UK first had standards for radio interference 30 years ago for the protection of radio broadcasting; today most household electrical equipment is suppressed by the manufacturer but a lot of equipment is not.

In hospitals the departments where interference may be a problem are ECG, EEG, EMG, theatre and the pathology laboratories. The difficulty with pathology laboratories is that equipment which creates interference is likely to be in the same room as equipment which is liable to be affected by interference. The apparatus which to date has been found to be the most susceptible to interference is the FN blood cell counter. It was therefore carefully examined in the laboratory and means were determined which would enable the counter to operate satisfactorily in the presence of interference; results were published in the journal *Hospital Service Engineering* (1); this journal is sent free to all Area Engineers in the health service. Additionally, the manufacturer agreed to adopt the modifications in future production and also to modify those already in use in hospitals.

The equipment which has caused most interference is the centrifuge, the sparking at the commutator being the cause, while auto-diluters have been a close second. The department has published appropriate means for suppressing centrifuges in *Hospital Service Engineering* (2).

Basically, interference is generated whenever a circuit is made or broken, whether by a commutator, thermostat contact or switch. The amount of interference created will vary from zero to a maximum depending on the point in the cycle of the alternating current at which the switching occurs. Interference suffered by an item of equipment is found in most cases to be due to another item of equipment in the same laboratory, and frequently it is on the same final subcircuit (30 A ring in the case of 13 A socket outlets).

Normally, interference is conducted from the equipment which produced it along the mains wiring to equipment which is susceptible to interference. In such cases it is easy to determine whether the British Standard permitted levels are being exceeded and if so, to put suppression filters in the mains supply lead. In some cases interference can be radiated from one item and picked up by another. This can occur in a pathology laboratory if the physiotherapy department is too close, in which case short-wave diathermy at 27 MHz will be the most likely cause. Corrective action cannot be taken at the source since radiation from the equipment to the patient is fundamental to the treatment. Therefore, the points at which the radio frequency signals are picked up must either be screened or the circuits fitted with radio frequency rejection components; the principles are well known, hence a cure is possible.

Problems are frequently met when one item of equipment generates signals which are meant to be accepted by another item which is not immediately adjacent to it. The techniques of transmitting an electrical signal along a cable and receiving an accurate reproduction of it at the far end are well known to those engaged in telecommunications. Unfortunately, many outside the telecommunication field imagine that the signal which is put into a cable will come out unaltered at the far end. A wire with a screen around it is not proof against interference at low frequencies. Fortunately, balanced amplifiers are readily available today and these can make up for many of the deficiencies of the interunit cabling. The technically correct method of transmitting signals from one unit to another is to generate a signal which is balanced relative to earth, to transmit it along a screened twisted pair cable and to receive it in a balanced amplifier. Methods using coaxial cable can succeed if the cable lengths are not too long, and signal levels are greater than 1 V and a balanced amplifier is used to receive the signal.

Some computer manufacturers demand that their equipment should be supplied with 'clean mains' and a 'noise-free' earth. In a practical hospital situation these do not exist; if the equipment cannot operate on normal mains then the manufacturer must fit interference suppression filters on the mains feed to his equipment. The ordinary mains voltage, while being basically a sine wave having an r.m.s. value of 240 V, has superimposed on it various harmonics of 50 Hz, switching surges, occasional impulses of only a few microseconds duration but up to 400 V in amplitude, regular impulses at 50 Hz and 100 Hz due to the operation of silicon controlled rectifiers and fluorescent lamps, and also radio frequency voltages and noise. Some of these unwanted signals can readily transfer themselves from a mains cable into cables carrying information signals; it is therefore important that signal cables of any length should be run in steel trunking, with mains cables in a separate steel trunking. In one hospital both the signal cables and the mains cables had been supplied considerably longer than was necessary and they were left coiled up together on the floor, thus, unfortunately, ensuring a maximum transfer of mains noise into the signal cables.

In 1965 the Ministry of Health published a memorandum (3) which dealt with some hospital interference problems and their solutions. Hospitals were advised that practical assistance in clearing interference in hospitals was available for the asking. Since that time considerable experience has been gained in solving actual problems and that experience is available to all hospitals in the Health Service.

REFERENCES

1. *Hospital Service Engineering,* November, 1972
2. *Hospital Service Engineering,* March, 1970
3. *Hospital Technical Memorandum No. 14,* Ministry of Health, 1965

ELECTRICAL INTERFERENCE IN CLINICAL LABORATORIES

F H Baker TEng (CEI) MITE

In 1965 the Ministry of Health, as it was then, advised hospital authorities, that because of the special nature of the problems involved, the Engineering Division would upon request, investigate specific problems of electrical and radio frequency interference to electromedical equipment.

Initially the investigations were concerned with interference to physiological recording equipment, ECG, EEG, etc, but since 1970 the number of requests to investigate problems in clinical laboratories has increased, and the object of this paper is to highlight the major points of a few of the investigations conducted.

The first major investigation followed a request in 1970 to investigate radio frequency interference to recording and computer equipment in a biochemistry department accommodated on the third floor of a teaching hospital in central London. It was reported that the cause of the interference was the shortwave medical diathermy equipment in use on the sixth floor of the building in the physiotherapy department, and initial tests quickly verified this report.

Shortwave physiotherapy diathermy equipment intentionally generates a 27.12 MHz carrier frequency, amplitude modulated with 50 or 100 Hz. The frequency 27.12 MHz is an internationally agreed frequency for industrial scientific and medical use and in the band 27.12 MHz ± 0.6 per cent unlimited radiation is allowed. The radiation of harmonics in the UK is controlled by legislation detailed in Statutory Instrument 1895. A Department of Health specification limits the depth of modulation on shortwave sets to less than 30 per cent and tests confirmed that the sets in use in the hospital complied with this specification.

The radio frequency field at 27 MHz measured inside the laboratory was above 100 dB rel 1 μV/m, and the radio frequency noise voltage at this frequency on the electrical supply was greater than 90 dB rel 1 μV.

The Department's Technical Memorandum No 14 recommends that the approximate minimum spacing between shortwave sets having 10 per cent modulation and equipment having 20 μV/cm sensitivity should be 60 feet (18.5 metres). In this case the shortwave sets were only 25 feet (7.5 metres) vertically above the laboratory and the radio frequency coupling between the two areas was improved by service pipes and ducts connecting the two areas.

The equipment in the biochemistry laboratory suffering the interference was used for the quantitive analysis of chemical solutions. The solutions were passed through an autoanalyser system consisting of a pressure pump, flowcell and colorimeter. Two output voltages, a reference voltage and a measurement voltage between 1 and 10 millivolts, depending on the solution under test, resulted.

Originally these voltages were transmitted direct to a potentiometer pen recorder where peak values were determined manually from the record and then processed. This system was not affected by the short-wave sets in the physiotherpay department.

The introduction of an on-line computer working to process the information necessitated the introduction of high gain amplifiers between the colorimeter output and the pen recorder. These amplifiers, one for the reference signal and one for the measurement signal amplified the colorimeter output signals to between 1 and 10 volts necessary for computer operation; and also provided a 1 to 10 millivolt signal to operate the pen recorder.

Following the introduction of the amplifiers severe interference occurred taking the form of a shift of up to 5 inches in the chart recorder baseline when shortwave physiotherapy sets were in use. It was considered that detection of the 27.12 MHz modulated signal was occurring in the amplifier, and laboratory tests were conducted on a spare amplifier to determine the method of entry of the radio frequencies into the amplifier. As a result of these tests numerous modifications to the amplifier circuitry and console unit were suggested and carried out by the manufacturer.

One interesting aspect was observed during the tests, the 27 MHz was being picked up by the amplifier recorder connecting cable and fed into the amplifier via the *output circuitry* and 1000 pF capacitors were necessary between the output terminal and common line.

In addition to modifications to the amplifier itself, the cable used for the transmission of signals from the colorimeter to the amplifier consisted of a screened twin pair. The screen was used as a common line for the return path of the reference and measurement signals. This was unsatisfactory and was replaced by a 4 core screened cable allowing the screen and common line to be separated.

A number of reports of interference to a blood cell counter provided an interesting investigation. The counter operated on the principle that particles of different size would cause the electrical resistance between two electrodes to change accordingly. Now it is strongly recommended by the Department that when purchasing new equipment, hospital authorities should ensure that if equipment is liable to generate radio frequency interference, the levels generated should not exceed the recommendations of BS 800 'Radio Interference Limits and Measurements' Parts 1-3 1972.

Additionally equipment susceptible to interference should work in an interference environment of 12 dB above the BS 800 recommendations.

The repeatability and accuracy of blood cell counts was being affected by external interference from other items of laboratory equipment, e.g. centrifuges, diluters, staining machines, even though the items were suppressed and generating levels of interference well below the BS 800 figure.

The standard remedy of fitting a radio frequency mains filter and screened flexible mains cable to the blood cell counter only improved the performance marginally. Subsequent tests in the laboratory proved that the counter was affected by signals only at certain frequencies. A high level of susceptibility occurred at 41 MHz with lower responses at 11.6, 15.6 and 49 MHz.

It was necessary therefore to prevent these higher signals from reaching the amplifier, and this was achieved by fitting a 4.7k resistor in the input circuit immediately following the aperature current circuit.

Following further tests by the manufacturer to ensure that the normal operation of the equipment was not affected, the modifications recommended were incorporated on new models, and hospitals were advised through Hospital Service Engineering so that existing models could be modified.

The majority of investigations of interference to computer equipment are complicated by the fact that the equipment is only affected on isolated occasions—once a day for example. The cause of an intermittent equipment malfunction is very difficult to determine, especially as the problem seldom arises when the hospital is visited. It is necessary therefore to investigate logically and I consider three questions need answering.

1. Is the equipment itself operating satisfactorily? It should not be assumed that all malfunctions result from outside interference. Two recent problems were satisfactorily cleared by ensuring that the tapes used in the equipment were clean and of the recommended type.

2. Is the wiring of the installation satisfactory? In too many cases the answer to this question is no. In a midlands hospital recently the mains and signal cables were entwined together in large loops on the floor behind the equipment. Another item of equipment had unscreened mains wiring and signal cables laced together. Problems are sure to arise when close coupling of this nature exists between mains and signal cables.

3. Is the malfunction due to mains supply abnormalities? We are assured by manufacturers that computer installations are sensitive to mains supply variations, although when pressed they are normally reluctant to specify figures and expressions such as 'a good mains supply is required' are quoted.

Supply abnormalities that have been encountered are:

a. *Transient impulses*—characteristically these impulses have a duration from less than 1 millisecond to a few seconds and waveforms from single, fast rising, exponentially decaying pulses to undamped oscillatory disturbances. Peak amplitudes vary from a few volts to a few hundred volts.

b. *Sags and surges*—rapid increases or decreases in the magnitude of sine wave that persist for a major part of a cycle or for more than one cycle of line frequency. A recent record showed a sag from 240 volts to 72 volts for a 4 cycle period.

c. *Line interruptions*—one tends to think of mains supply failures lasting for a few minutes, but interruptions for one or two seconds do occur. During a recent investigation at a hospital in the South of England, the mains failed for 3 seconds, long enough for the computer equipment to malfunction.

d. *Frequency variations*—these are unlikely to be a problem. Deviations of more than 0.5 Hz are rare.

During investigations where supply abnormalities are suspected the need to monitor the mains supply to the equipment and to relate any variations to time is essential as the variations occur intermittently. It is also essential that during the same period an accurate time record of equipment malfunctions is kept. Information derived from supply mains monitoring can also indicate, assuming no supply abnormality occurs during an equipment malfunction, an intermittent fault in the equipment itself. Commercial monitors are now available which will analyse one and three phase mains supply systems and classify each disturbance into transient impulse, cycle to cycle change, or slow RMS change. When preset thresholds are exceeded, the instrument prints out time, classification, magnitude, duration and channel designation. Two versions of this type of equipment are known to be available, and cost approximately £4,000.

ON-LINE COMMUNICATION BETWEEN THE LABORATORY AND THE CENTRAL HOSPITAL PROCESSOR

Professor K Borner, H Enke, W Lange and G Maigatter
Institut fuer Klinische Chemie und Klinische Biochemie,
Freie Universität Berlin, Klinikum Steglitz
Hindenburgdamm 30 D-1000 Berlin 45

In 1974 the administration of our hospital planned to replace the outdated central processor. Many staff members of the hospital needed extended computer capacity. The users of small computers especially felt the limitations of their equipment. The Chemistry Laboratory—which produces approximately one million specified test results per year—used at that time both a small process computer for data acquisition and the service of the hospital's computer centre for handling bulk data. We wanted to overcome the obvious limitations of this configuration by means of teleprocessing mainly for the following reasons:

1. We wished to make timely use of the patient admission and identity data held on the central computer and furnished by the admitting office.
2. We wanted to shorten the time interval required for exchange between laboratory and computing centre from hours to minutes.
3. We wanted to optimise the use of the different properties of both computers.
4. We were urged to make an economic use of centralised equipment.

THE LINKING OF SMALL COMPUTERS TO A NETWORK

The crucial question was how to incorporate minicomputers of various types and sizes into the network. We wanted a technical solution that fulfilled the following conditions:

1. Use of standard hardware throughout,
2. A uniform communication procedure independent of the individual manufacturers with ready-to-use software packages for remote data transmission,
3. A maintenance guarantee for both hardware and software, defined as an essential part of the standard contract (BVB).

After discussions with several manufacturers we decided to use the transmission procedure for the IBM 2780 terminal (1) based on the IBM protocol for Binary Synchronous Communications (2). At that time we were assured by at least four manufacturers that suitable software was available. In short this is a procedure for the transmission of command and of data files.

HARDWARE REQUIREMENTS

For the purpose of remote data transmission three groups of devices are required in a central data network:

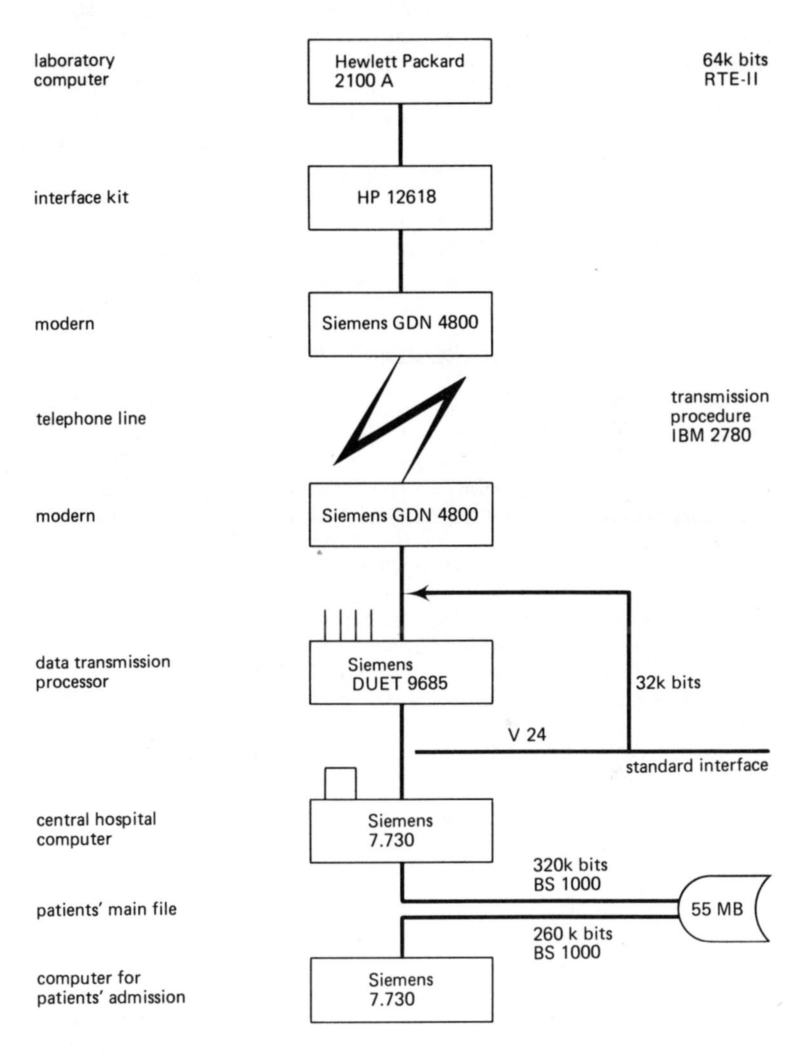

Figure 1. Linkage of the Laboratory Computer to the Central Hospital computer

1. Interfaces for the satellite computers (e.g. synchronous data sets),
2. A system of communication lines including converters (e.g. modems),
3. Interfaces for the central processor (e.g. a front end processor).

In our case (Figure 1) the central processor (Siemens 7.730) was extended by a dedicated front end processor (DUET 9685) with five I/O buffers for satellite computers. Normal telephone type lines are used for transmission but these have been isolated from the voice telephone network to prevent crosstalk interference. Data conversion is performed by modems (GDN 4800). Additional lines for voice co-ordination between the local and central operator are strongly recommended. Both processors, local and remote, required considerable extensions of main memory for the storage of program modules for teleprocessing. In our particular case all hardware for the laboratory computer (Hewlett-Packard 2100A) was furnished by the manufacturer in the form of a package (HP 91780 Remote Job Entry package).

SOFTWARE REQUIREMENTS

The operating system of the central computer (BS 1000) provides additional modules for remote data transmission and remote job entry and a dedicated operating system for the front end processor. Of the existing user-written programs only the I/O routines had to be adapted to remote data transmission. Software for the laboratory computer was acquired from the manufacturer as part of the above quoted package—also called RDTS: remote data transmission system (3, 4). By means of this system the local computer can still be run in the multiprogramming mode while handling remote data transmission. In our application, all laboratory data acquisition programs continue during communication with the central processor. Consequently, there are quite a few restrictions upon the internal organisation of the laboratory work. Besides the generation of new files and job streams we had to write only the programs to convert the internal record and data formats into a code suitable for data transmission.

ESTIMATED COST OF THE COMPUTER TO COMPUTER LINK

The exact determination of the cost for a subset of functions such as teleprocessing, within a major computer project is rather difficult. Our cost must be considered in relation to the assumption that five small computers will be finally linked to the central processor. Our total costs for the interfacing amount to about 95,000 DM at 1975 and 1976 prices. This figure does not include the cost of providing a telephone line nor the labour costs of our own computer staff.

EXPERIENCE DURING THE INSTALLATION OF THE LINK

It took about 4 months to get the hardware working and to implement the software with generous support by both companies. We successfully tested the connection under various conditions. Apart from a few minor hardware faults we were able to operate the system with satisfactory reliability. We therefore commenced routine use of the system in August

1976, and now we have only a few remaining problems in the software (e.g. unexpected page ejects of the line printer). A major difficulty in locating and correcting these errors lies in the fact that source listings of such expensive software are difficult to obtain by the customer.

Attempts to connect the laboratory computer to a CDC 27801 terminal were also successful.

EXPERIENCE WITH REMOTE JOB ENTRY UNDER ROUTINE CONDITIONS

We have now worked with remote job entry (RJE) for more than 12 months under routine conditions. The laboratory computer is handled by two data typists for most of the time. A typical sequence of events including the use of the RJE facility is as follows:
The daily production of some 3000 test records is entered into the local computer and temporarily stored. After an editing procedure the data are converted and transmitted to the central processor. Transmission includes a command file which initiates a job sequence in the remote computer. Among other jobs the laboratory data are matched up with the appropriate patients. After an interval the local operator schedules another job sequence whereby the completed laboratory data are received from the central processor and are printed out on the line printer. Similar sequences serve the exchange of patients' main data. The example shows the co-operation of both computers with partial overlap of functions in order to guarantee backup. Briefly summarised, the local computer in the laboratory has the following functions:

1. Acquisition of test results, checks for plausibility,
2. Intermediate storage of results and patients' data,
3. Editing procedure,
4. Printing of reports (optional),
5. Internal retrieval of results,
6. Transmission of laboratory results and reception of patients' data.

The central (remote) processor completes this catalogue by the following functions:

7. Intermediate storage of laboratory data,
8. Provision of data on the patients,
9. (Interpretation of results–in preparation),
10. Generation of reports,
11. Transmission of reports,
12. Final storage of laboratory data,
13. Special evaluations.

CONCLUSION

Linking two computers together is still somewhat expensive. Nevertheless it can be made to work sufficiently reliably even with computers of various size and origin. Remote job entry (RJE) might become a reasonable alternative for people who cannot afford a 'complete' laboratory computer system.

REFERENCES

1. Data Transmission Terminal–Component Description. Form A27-3005-2, IBM 1867, IBM 9780, devised August, 1970.
2. General Information–Binary Synchronous Communications IBM Form 6A-97-3004
3. Remote Data Transmission Subsystem. General Information Manual. Hewlett-Packard Form 91780-93002
4. Remote Data Transmission Subsystem. Programming and Operating Manual. Hewlett-Packard Part No. 91780-93001

Part 6

COMPUTER ASSISTED CHOICE AND USE OF LABORATORY RESULTS

'DAMMING THE FLOOD OF LABORATORY DATA'

Arthur E Rappoport, MD, **Yuel D Tom**, MD,
William D Gennaro, MS, and **Robert E Berquist**, BEng

INTRODUCTION

The Youngstown Hospital Association (YHA) comprises the North Unit (NU) possessing approximately 600 beds located five miles distant from the South Unit (SU) with 350 beds and a 100 bed babies and children's hospital (TBC) adjacent to the NU. Each operates an active ambulatory care facility (ACF). The hospital and its 300-member medical staff are components of the clinical faculty of the Northeast Ohio Universities College of Medicine (NEOUCOM).

LABORATORY CENTRALISATION

A full service clinical laboratory to serve the entire institution has been established at the NU while a relatively small, but adequately equipped and staffed, satellite 'Stat-Lab' is maintained in the SU to perform urgently required tests and to procure all specimens to be transported by frequent courier to the NU for routine testing. A total of two and a half million tests were performed in the division in 1976.

CONSOLIDATED ANALYTIC SUBSYSTEM

As a result of these steps and extensive use of automated, multichannel, testing instruments, large batches of specimens could be analysed rapidly by a wide variety of procedures, particularly in chemistry, hematology and serology. These developments have contributed to significant reduction of laboratory cost through elimination of all possible duplication of staff, space and equipment. Enhanced supervision of test performance could be maintained constantly because of establishment of standardised quality assurance practices and the availability of senior consulting staff at one place.

Approximately 95 per cent of the total SU routine chemistry testing was transferred to the NU after acquisition of an SMA/6 and 12. About 80 per cent of the SU hematology examinations were also shifted to NU after procurement of a Technicon SMA/4 and which subsequently was replaced by a Coulter 'Sr'. The SMA/4 was then modified to perform automated platelet, white cell and hemoglobin tests especially used in the oncology profile. Installation of a DADE Auto-Fi at NU permitted almost total centralisation of prothrombin analyses while purchase of a Packard Auto-Gamma counter led to a localisation of all RIA examinations at NU. Blood banking (donor recruitment and procurement and specimen drawing and processing) including serology and HbSAG was retained at SU, while appropriate blood inventory for the transfusion service was maintained daily at NU. 'Stat' Microbiology and routine urinalysis were retained at SU for obvious reasons. Virology, immunology, electrophoresis, cytogenetics, toxicology, and other complex procedures

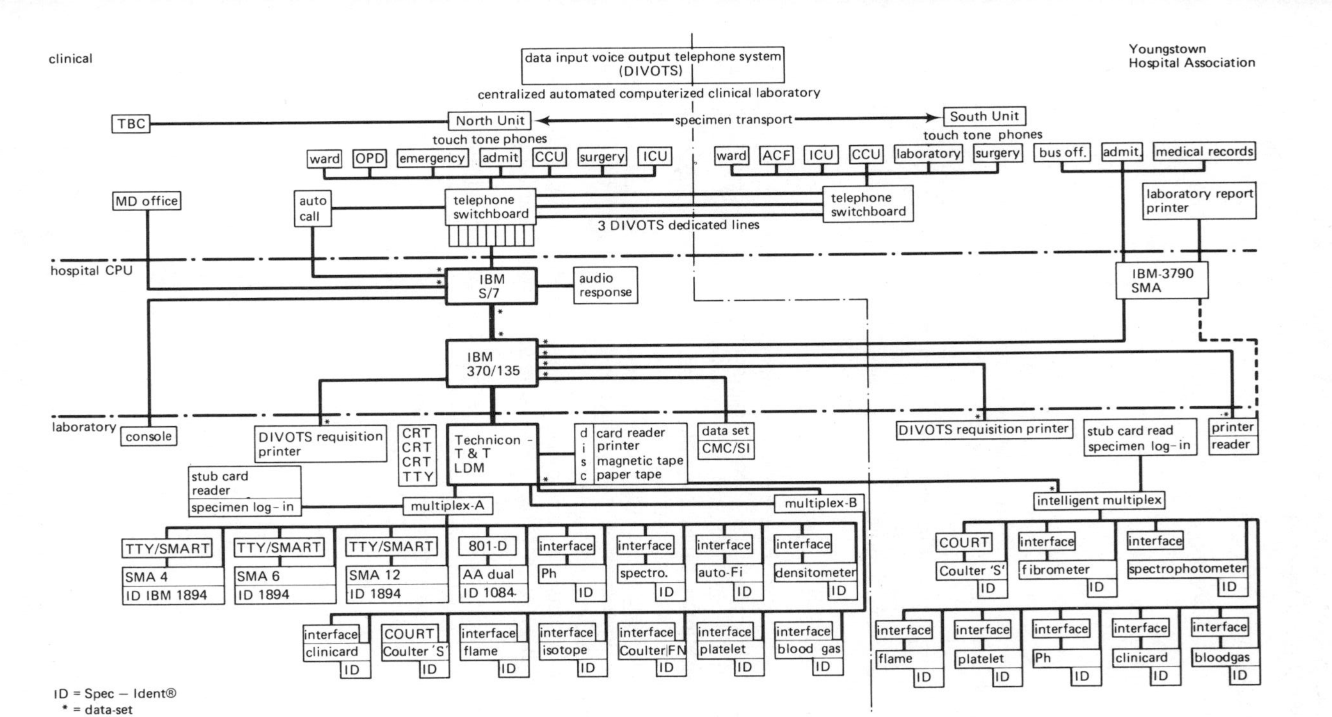

Figure 1. Flow Chart showing the total laboratory configuration of DAS, LIS and LCS. Note the telephone communication from all touch-tone phones throughout the hospital to the IBM S/7 through switchboards at both North and South Units.

Requisition printers are located in both unit laboratories. Note the analytic instruments, special interfaces and Spec-Idents connected on-line to the LDM through a specially designed, intelligent, digital submultiplexor possessing floppy discs to log in all data.

The LDM performs data reduction, specimen identification, result calculation and automatic on-line transmission to the 370 of results which then become immediately available for telephone communication to physicians through DIVOTS.

became completely localised at the NU, along with microbiology and surgical pathology (Figure 1).

LABORATORY INFORMATION SUBSYSTEM (LIS)

Data Acquisition Subsystem (DAS)

One of us (AER), in 1963 as consultant to IBM, participated in an early pioneering project to create a computerised LIS. One part of this effort consisted in the design of an off-line, batch DAS to include automatic test order entry, test data entry and result calculation and reporting. Most of the then available automated (AA) and semiautomated electronic instruments (spectrophotometer) communicated directly to the DAS device through retransmitting potentiometers or other interfaces. Peak-picking, peak-holding, and A/D conversion yielded raw data values which were punched automatically on cards and read into the hospital computer (CPU) where the test results were edited and released by the technologists into the patient's master file (PMF) from which daily ward and updated cumulative summary reports were generated and sent to the patients' wards.

Automatic Specimen Identification (Spec-Ident)

A second activity was the creation of the first operational, human and machine readable, specimen identification (ID) system. This employed small, sequentially prenumbered and prepunched stub cards which were attached to specimen tubes by phlebotomists after venepuncture. The specimen number was merged with the patient's hospital number through appropriate software. The stub cards were read in specially designed readers on AA turntables or in modified badge card readers (Spec-Idents) coupled to the analytic instrument.

Manual test results, i.e. urinalysis, differential blood cell counts, etc., were entered directly into the CPU through key-punchcards or port-A-punchcards.

This initial DAS was expanded by acquisition of more efficient hardware and communication devices and by improved software to support the professional and administrative functions. The original CPU (IBM 1401) was replaced by a 360/30, then by a model 40, which gave way to a 370/135, and the present to 370/148.

Dedicated On-Line Laboratory Computer

Responding to the rapidly increasing test load, plans were initiated eight years ago to transform this off-line LIS to an on-line mode by procurement of a dedicated laboratory computer. We continued to share the hospital's CPU to enjoy the benefits of its great power, rapid internal processing rate, large storage capacity and comprehensive patient data base. This decision led to the acquisition of the LDM, manufactured by the T & T Corporation, which subsequently has been acquired by the Technicon Corporation.

The LDM is structured around a Data General, Nova Model 2 Computer plus many peripherals which include several cathode ray tubes (CRT), a printer, card reader, magnetic and punched paper tape readers, and several multiplexors to which all analytic instruments are interfaced. Its basic function is to perform data acquisition, data reduction, organisation of data and transmission of results in a bidirectional mode to the host computer which is accomplished by a high-speed data set (WE 4800). Thus, it possesses three distinct subsystems:

a. Autonomous data acquisition modules for the analytic instruments in the laboratory.

b. Systems communication modules for communication between the 370 and LDM.

c. The mainframe, consisting of CPU, core memory and system I/O controller.

Efficient software has been developed to create a DAS for most of the conventional automated and semiautomated electronic instruments as well as to accommodate manual data inputs through CRTs, key-

Table 1 Sixteen NU laboratory instruments and Sixteen Spec-Idents coupled on-line to LDM through multiplexors A and B (see Figure 1) and capable of simultaneous input to the LDM

Instrument	*Tests performed*	*No. of tests*
Spec-Ident	to log-in specimen arrival time in laboratory	All specimens
Coulter S 'Sr.'	hematology: RBC, WBC, HGB, PCV, MCV, MCH, MCHC	7
SMA/4	platelet, WBC, hemoglobin	3
SMA/6	glucose, BUN, K, Na, Cl, CO_2	6
SMA/12	total and conjugated bilirubin, Ca, Phos, alkaline, phosphatase, SGOT, SGPT, LDH, CPK, uric acid, creatinine, total protein	12
autoanalysers	cholesterol, triglyceride, PKU, acid phosphatase, HDL-LDL	6
pH	pH	1
spectrophotometer	iron, lipase, amylase, microbilirubin, plus all other conventional colorimetric tests measured by %T or absorbance	20
auto-fi	PTT, pro-time, fibrinogen, coagulation factors (automated)	6
Helena quick scan flur-vis	protein, lipoprotein and hemoglobin electrophoresis, isoenzyme of LDH, CPK and alkaline phosphatase	12
clini-card	'stat', sugar, BUN, Ca, bilirubin, salicylate and others	22
bloodgas (Corning)	PO_2, PCO_2, bicarbonate, pH, total CO_2, base excess	6
isotope (Packard)	HbSAG, T_3, T_4, renin insulin hormones, blood volume, etc.	6
Coulter ZBI	'stat', RBC, WBC	2
Coulter platelet counter	'stat', platelets	1
Perkin-Elmer (Hitachi) spectrophotometer	kinetic u.v. enzyme analysis, SGOT, SGPT, LDH, CPK, HBD, GGT, EMIT, toxicology	6

punchcards, or port-A-punchcards, typewriter consoles, punched paper or magnetic tape, and magnetic cards. Thus, it was possible to merge the host's stored demographic data and the laboratory data files for a total LIS serving all patients in all units.

Table 1 lists those NU instruments currently operating on-line to the LDM. Each is equipped with an appropriate interface and a Spec-Ident (machine readable specimen identity).

Shared Data Acquisition System at SU

Through installation of an additional digital submultiplexor (DSM) in the SU laboratory, many of its semiautomated (Clinicard) and automated (Coulter 'Sr.') instruments associated interfaces and Spec-Idents were connected by data sets to the NU LDM. These SU results were stored automatically in the patients' 370 files and merged with the same patients' results generated in the NU centralised laboratory. Table 2 lists those SU instruments and their associated devices which are connected to the DSM.

Table 2 SU laboratory instruments, interfaces and Spec-Idents coupled by DSM to the LDM at NU

Instrument	*Tests performed*	*No. of tests*
Coulter Sr	hematology: RBC, WBC, HGB, PCV, MCV, MCH, MCHC	7
Coulter-FN	platelets	1
clini-card	'stat', sugar, BUN, Ca, bilirubin, salicylate and others	20
Corning gas	PO_2, PCO_2, bicarbonate, pH, total CO_2, base excess	5
fibrometer	pro-time, PTT	2
spectrophotometer (Coleman)	iron, lipase, amylase, microbilirubin plus many other conventional colorimetric tests measured by %T or absorbance	20
flame photometer (IL)	Na, K, Li	3
Cl-CO_2-Corning	Cl, CO_2	2
pH – (IL)	pH	1

Keypunching, card reading capability and CRTs are available at the SU to report and transmit manual test data. As a consequence of this configuration, the satellite laboratory is completely capable of batch and on-line processing of all locally produced results by the NU dedicated laboratory computer. From the foregoing, it should be obvious that the LDM, in addition to serving as a data reduction device, also acts as a communication terminal since it controls communication between the two units' laboratories and transmits all laboratory data to the hospital CPU.

THE HOSPITAL CENTRAL COMPUTER (CPU)

YHA possesses a large CPU which, through distributive processing, is available to all units and departments of the hospital for professional and administrative data processing. Physically located at the NU, it services all units through dedicated telephone communications lines and data sets to various terminals. These include an IBM minicomputer (3790) at the SU, capable of printing the laboratory ward and cumulative patient summary reports as well as handling all other medical and management data processing for that unit.

In the SU laboratory, IBM keyboard, printer and card reader terminals are employed to transit results directly to the 370 and to receive reports abstracted from patients' files. Several CRTs are distributed in the admitting office, business office and outpatient central registry of both units. The LDM, acting as a terminal, also transmits to and receives data from the 370 at 4800 baud. In addition to CRTs and cards, manual data also may be entered either into the LDM or 370 by key-to-disc machines ('floppy disk'). The diskettes are reusable and each can store approximately 1800 card images. Communicating magnetic card/selectric typewriters (CMC/ST) are also used to transmit free, unstructured text directly to the 370 for entry into the patients' files.

All on-line message switching is handled by IBM customer information control system (CICS). Reports are printed at the NU at the rate of 1100 lines per minute. Programs on the 370 have been written in Cobol.

Purchase of a software package, MARK IV, (Informatics Corporation, Inc.) provides greater programmer productivity for all areas of the Youngstown medical information processing system (YIPS).

LABORATORY TEST DESCRIPTION FILE

Eight hundred laboratory procedures are listed and described in the 370 test description file. Each procedure is identified by a 4-digit number including a self-check digit and linked to the range of normality tables adjusted for the patient's age and sex. All procedures are catalogued in a publication, 'DIVOTSrectory', (see below) which is distributed to all Nursing Stations and physicians' offices. Tests listed alphabetically, are cross-referenced to synonyms and also are grouped in profiles according to a unique YHA program, problem assisted laboratory medicine (PALM, see below).

COMMUNICATION SYSTEM FOR ORDER ENTRY (O/E) AND RESULT RETRIEVAL (R/R)

These long neglected vital steps in the total test path have undergone extensive mechanisation to accompany similar developments in the analytic and documentation subsystems. In the conventional test requisition (O/E) systems, multipart, preprinted, color-coded paper requisitions are usually filled in by nursing station personnel using embossing devices (addressograph) to enter patient demographic data and by checking boxes to indicate the desired tests.

When delivered to the laboratory, the various data are entered into the computer by usual EDP techniques. Although this process reflects a significantly improved step over earlier manual methods, substantial O/E communication gaps continued to exist between nursing station and laboratory and R/R steps from laboratory to nursing station. Similarly, recovery of urgently required results prior to preparation and delivery of printed reports usually was a time-consuming, error-filled, tediously difficult, process.

After reviewing and evaluating many of the available systems recommended for on-line communication, including ward based CRTs (with and without light pens), dual-button CRT displays and conventional printers, we considered the use of touch-tone (T-T) audio-response telephone technology as a possible method of eliminating time and space in the test cycle from physician's request to the laboratory and result reporting. This represented an attractive option because of its relatively low cost, reliability and easily mastered technology in addition to the obvious advantage of utilising the existing universally available, and already paid for, hospital telephone system which is maintained 24 hours a day by the local phone company.

DIVOTS (data input voice output telephone system) for Order Entry (O/E)

O/E: Requests for laboratory tests from both unit Nursing Stations or physician's offices are initiated by entering the test code through T-T phones (or Rotary phones supplemented by a T-T pad) connected by Data Sets to an IBM System/7 computer located in the NU laboratory. The S/7 possesses an audio response disc containing an 850 word, laboratory orientated vocabulary, including all numbers and letters, Greek symbols, test names, administrative and technical words and all possibly anticipated descriptive results from all laboratory divisions including haematology, urinalysis, chemistry, immunology plus many terms in microbiology, virology and anatomic pathology. Polysyllabic, compound words can be created and uttered, utilising available prefixes (micro, hypo) and suffixes (cytes, osis) coupled to the stem. The vocabulary was dictated to tape from which digitised phonograms were made and stored on disc. Tests are ordered by entering the patient's hospital number and test code by T-T (Figure 2) phones into the S/7, which immediately responds by spelling the patient's name and stating the names of the tests ordered. Errors may be deleted by entering the asterisk(*). Special conditions are identified by special codes, i.e. 'to be done stat or next day or specific time', etc. On completion of O/E, a requisition, listing desired tests and patient information, bearing six machine and human readable, stub cards is printed immediately within the laboratory (Figure 3). The stub cards are used by phlebotomists to obtain and identify the required samples. Upon returning to the laboratory, the phlebotomist logs her own and the sample ID numbers into the LDM, utilising a special reader, the Spec-Ident (Figure 4). After performance of the test, the results processed in the LDM are edited for accuracy and precision by the technologist who approves and LDM automatically transmits them to the patient's 370 file. O/E at the SU is performed in the same fashion utilising dedicated telephone trunk lines between the two units' switchboards (Figure 1) and thus communicates with the 370/S/7.

Figure 2. The use of a touch-tone pad, card dialer accessory attached to a conventional rotary telephone to input codes into the 370.

DIVOTS for Result Retrieval (R/R)

To obtain results rapidly and as soon as possible, physicians and nurses initiate DIVOTS by a similar technique but employ a different service (request) code. The patient and test are identified in like fashion and, to insure confidentiality, the physician by his personal ID number. DIVOTS spells the patient's name, states the name of the test and date and presents the results, indicating whether it is high or low abnormal. If the result is not yet available, the current status of the sample is given.

DIVOTS Auto-Call: 'Don't Call Us—We'll Call You'

In special cases involving urgently required information or life endangering situations ('Panic reports'), DIVOTS contributes in an excitingly unique fashion to expedite reporting. Auto-Call, without requiring human participation, automatically relays specified categories of laboratory tests to the wards. Included are results for patients in various acute care locations in both Units (Coronary and Intensive Care Units); certain substances (abnormal values of glucose, potassium or hemoglobin, etc.); drugs (digitoxin) or type of request ('stat' preoperative etc.). After such data have been transmitted from the laboratory to the 370, that computer automatically dials the patient's nursing station or clinic telephone (or his physician's office) sounds the 'Fate' (Victory) theme from the Beethoven

Figure 3. This demonstrates a punched Dial-a-Card (lower right) and patient identification wrist band (top) bearing pressure labels imprinted with his hospital number. Note the DIVOTS requisition (middle) printed in the laboratory by the 370. The left-hand portion is the audit trail file copy displaying vital demographic and diagnostic data obtained from the PMF in the 370. Note the repeated printing of the specimen number (039379) under the preprinted number by the computer which merges it with the hospital number (8230633). The numbered punched stub cards, when attached to the tubes and read by the Spec-Ident (R), automatically identify the patient from whom the specimen was obtained. Note the location of the wrist band pressure labels above the computer printed number on the stub card to verify the patient's identification.

A duplicate requisition has been burst and the variously identified (SMA 12, 6, 4, etc.) stub cards have been attached to the tubes containing specimens to be tested for the procedures indicated on each stub card and file copy.

There are two extra labels on the stub card '4' (EDTA tube for hematology). One label will be attached to the slide to be stained for the differential count and the other will be attached to a special differential, Port-A-Punch card for reporting the result. Both labels serve to insure correct specimen and patient identification which is possible even in manual testing.

5th Symphony to identify Auto-Call. DIVOTS spells the patient's name, states the test and furnishes the urgently desired result. The advantages of this service are obvious. Since the Auto-Call response is logged by the S/7, proof is established that the information was received by a responsible individual at that ward at a specific time.

DIVOTS provides O/E and R/R through TT phones at any time from any place. Physicians may employ DIVOTS from their offices, homes or autos for both functions to expedite diagnosis and treatment and thus improve medical care. DIVOTS has communicated successfully with Youngstown from many places in Europe and the Orient by conventional long-distance service.

Figure 1 is a panoramic Flow Chart which displays the integrated organisation of the laboratories, its DAS, LIS and Laboratory Communication System (LCS) with DIVOTS serving for O/E and R/R.

It is obvious that this configuration operates according to several shared computer communication modes.

1. A shared DAS for both NU and SU laboratories through a single LDM.
2. A shared LIS through the linkage between the LDM and 370
3. A shared LCS through the coupling of the S/7 and 370
4. A linkage between the S/7 and telephone system using audio-response technology.

PROBLEM ASSISTED LABORATORY MEDICINE (PALM)

We have developed a subsystem to LIS entitled PALM. This is based on the assumption that most procedures have major relevance to specific organ systems, disease syndromes, special therapeutic situations or

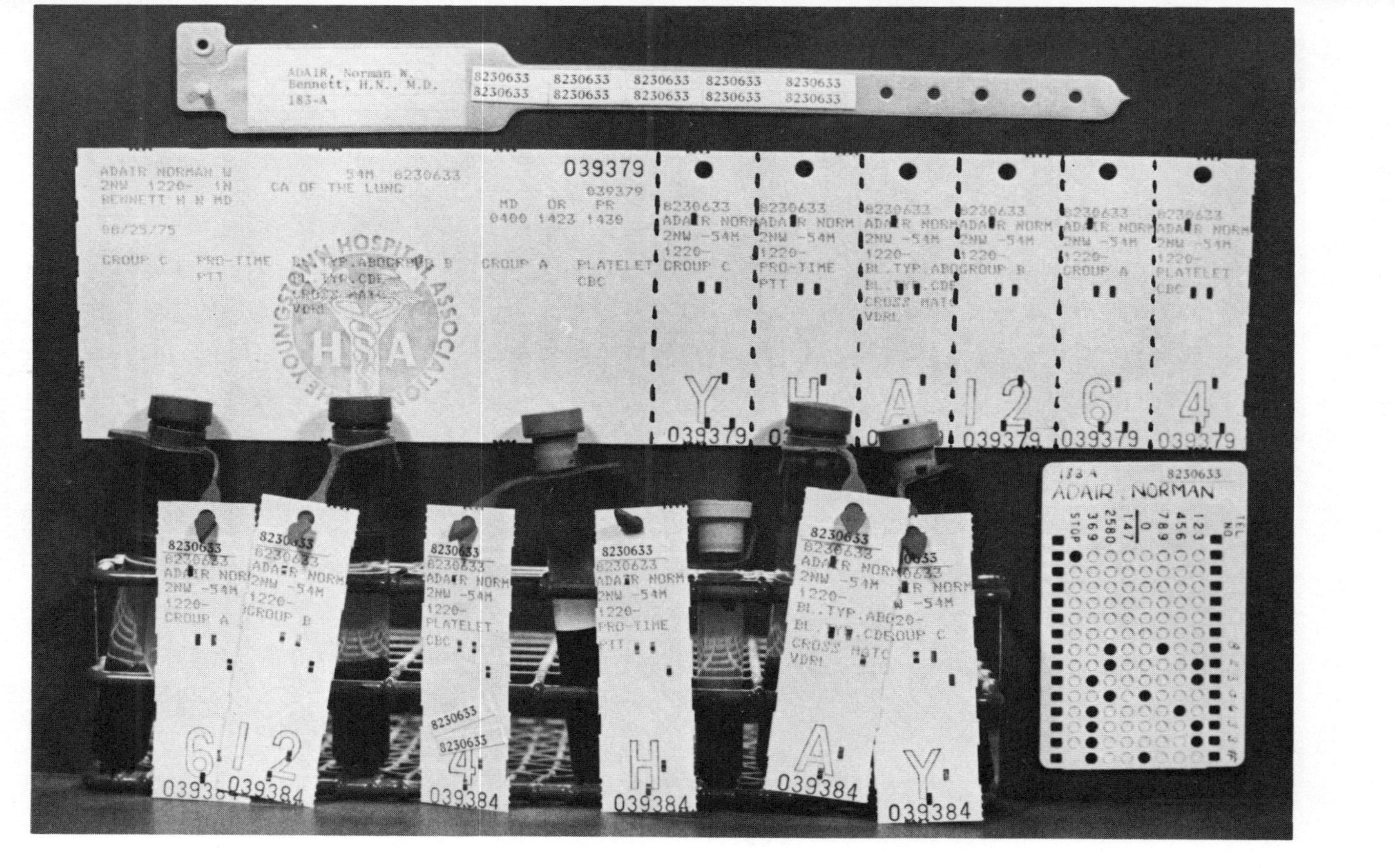
ADAIR, Norman W.
Bennett, H.N., M.D.
183-A
8230633
ADAIR NORMAN W 54M 8230633
2NW 1220- 1N CA OF THE LUNG
BENNETT H N MD
08/25/75
039379
MD OR PR
0400 1423 1430
GROUP C
PRO-TIME
PTT
BL.TYP.ABO
BL.TYP.CDE
CROSS MATC
VDRL
GROUP B
GROUP A
PLATELET
CBC
THE YOUNGSTOWN HOSPITAL ASSOCIATION
ADAIR NORM
2NW -54M
1220-
039384
183 A
ADAIR, NORMAN
TEL NO
123
456
789
0
147
2580
369
STOP

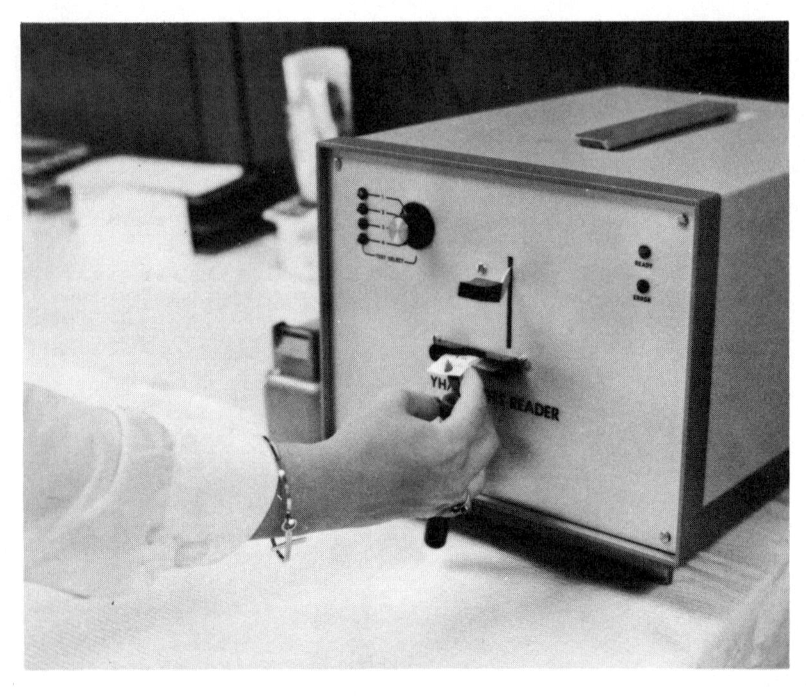

Figure 4. The stub card attached to the blood sample is inserted into a Spec-Ident (R) which reads the prepunched specimen number, sends it to the LDM and merges it with the test result prior to transmission to the 370. In this illustration, the sample number, its arrival time and the phlebotomist's identification is being logged. All semiautomated testing instruments possess similar identification units.

laboratory techniques. Accordingly, we have created a number of PALM profiles as follows:

Code Number	
9431	cardiovascular-hypertension profile
6971	coagulation profile
6645	dialysis (pre) profile
6653	dialysis (post) profile
5746	drug profile
9407	electrolyte profile
4936	GI profile
5886	hematology profile
6890	hormone, sex (male/female) profile
7234	hormone, pituitary profile
7269	hormone, adrenal profile
4375	immuno-hematology profile
6718	immuno-serology profile

9776	immuno-microbiology profile
5029	kidney profile
6041	liver profile
6874	respiratory profile
6777	spinal fluid profile
6858	sugar profile
6076	thyroid profile
6793	tumor profile

Within each Palm profile we have included those tests which we believe are most useful and relevant. Note the following examples: cardiovascular-hypertension and liver function:

9431 *cardiovascular-hypertension profile*	6041 *liver function profile*
SGOT	bilirubin total
creatine phosphokinase	bilirubin conjugated
CPK isoenzymes	alkaline phosphatase
LDH	SGOT
LDH isoenzymes	SGPT
cholesterol	GGT
triglyceride	hepatitis B surface antigen
lipid phenotype	bilirubin urine
digoxin	urobilinogen
digitoxin	protein, total serum
quinidine	protein electrophoresis serum
pronestyl-procainamide	protein electrophoresis pattern
catecholamine	urobilin
metanephrine	porphobilinogen
VMA	porphyrin-cop.
5-HIAA	ammonia
aldosterone	aldolase
bloodgas (cardiac cathet)	isocitric dehydrogenase
hydroxybutyne dehydrogenase	leucine amino pep.
lipoprotein pattern	ceruloplasmin
HDL (high density lipoprotein)	anti-mitochondria
LDL (low density lipprotein)	anti-smooth muscle
renin random	cholinesterase
renin fasting	alpha-feto-protein
renin postural	porphyrin-uro.
renin renal cathet	delta-alanine
myoglobin-urine	BSP
anti-myocardium	electrophoresis urine

O/E

Software has been developed to allow qualified medical personnel to order by DIVOTS a partial (screening) or total PALM profile by a single number. Each cluster includes selected procedures in hematology, chemistry, urinalysis, microbiology, immunology, RIA or from the entire gamut of available clinical laboratory examinations. These clusters are independent of the instrument technique or scientific discipline employed to generate the result. DIVOTS acknowledges audibly the entry of the profile, thus validating the correctness of the order and generates an appropriate requisition as described above.

NOVEMBER 20, 1975 DIVOTS ACTIVITY BY DOCTOR PAGE 4

	DOCTOR	NUMBER	PATIENT	WARD	ORDERS TEST NAME		ORDERS TIME	INQUIRIES TEST NAME	INQUIRIES TIME
20.	GARCIA ARMAND MD	8280762	ALLEY BETTY L	2W	CBC		14/46/46		
					LEE WHITE	6	14/32/32		
							14/32/32		
21.	GELRMAN F MD	8286744	MALONEY MARIE G	1E	CBC		09/54/15		
					MAS BL CUL		09/54/15		
					IRON STUDY		09/54/15		
					GROUP C	5	09/54/15		
22.	GEORDAN A W MD	8288364	WINING THELMA LOIS	2W	POTASSIUM	8	20/57/11		
23.	GONZALEZ J I MD	8283303	BORDYNIUK JOHANNA	3E	URIC ACID	5	16/05/47	URIC ACID	16/04/22
		8265802	OVENS FRANK J	2W	PCV	5	16/51/37	PCV	15/35/29
					HGB	5	16/51/37	HGB	15/33/24
								HGB	15/34/14
		8288577	YOUNG RICHARD H	1N	PTT		16/56/02		
24.	GREEN L N MD	8286051	HALSTEAD MARGARET E	1N	GLUC.TOL.	5	15/56/37		
		8288542	WITHERSPOON DOROTHEA E	4NW	CALCIUM	5	08/46/26		
					PHOSPHOR.	5	08/46/26		
25.	GUJU J G MD	8286086	RAMOS RENIE A	MAT				CBC	10/38/59
26.	HO P MD	8281610	BAILING EMERSON	2N	PRO-TIME	0	14/46/49	PRO-TIME	14/35/28
					PRO-TIME	5	14/40/47		
					PRO-TIME	0	16/12/44		
27.	HOFFMASTER R MD	8289182	ROMEO DOROTHY	251*	CBC	5	15/30/41		
					PTT	5	15/30/41		
					VDRL	5	15/30/41		

Figure 5. Record of DIVOTS activity. Notice the logging of enquiries in the right-hand columns.

DIVOTS thus consolidates multiple and diverse requests for tests to be performed in various laboratory divisions on a single requisition with 6 separate human and machine-readable, numbered stub cards. A phlebotomist can perform a single venepuncture efficiently and collect several optimally sized samples in several tubes with appropriate anticoagulants at one time. This eliminates duplication of requisition preparation, multiple venepunctures and excessive blood withdrawal while expediting sample procurement and alleviating patient discomfort. Physicians, employing PALM profiles, possess a checklist to consider the clinical appropriateness of available tests and thus can organise their studies more efficiently at the earliest possible moment.

As described above, test results are transmitted to the PMF in the 370 after test performance and quality control editing. Early transmission of test requests leads to prompter response by the laboratory and availability of data to the physician, who can make decisions concerning further treatment or diagnostic studies. In the latter case, he may request additional diagnostic studies that day. We are expanding DIVOTS to include the capability to order X-rays, EKGs and other diagnostic studies as well as drugs and other services.

R/R by DIVOTS

After initiating the TT request, the particularly desired PALM profile code is entered. This leads to a recital of all results within that profile or, if desired, only abnormal results. If none is present, PALM will so state.

PALM profile results are also available through Auto-Call. Such early availability of results leads not only to better medical care, but can materially shorten the turn-around-time of the test cycle, reduce hospital stay and permit considerable reduction in health care costs.

Figure 5 is a partial listing of the daily record of all O/E and R/R activity accomplished through DIVOTS according to physician, patient, test, and time of the transaction. It furnishes a log of management information of considerable importance to oversee the operation of the department.

PRINTED REPORTS

Many printed reports are prepared internally within the laboratory by the LDM for use by technical personnel, i.e. quality control, file copies of results, lists of abnormals, records of O/E and R/R transactions, log-in times of orders, specimen receipt, statistical analyses of test volumes, charges, fiscal and personnel management data which for reasons of space are not shown here.

Reports to physicians of test results are vital and two types are printed daily.

Interim Ward Report

This is printed simultaneously at both units and lists all test results performed between 6.00 p.m. the previous evening and that day up to 12.00 noon. This consolidates all results from all patients arranged by nursing station, room number and bed. It is delivered by laboratory personnel, posted at the nurses desk and is available for review by physicians and nurses.

** RETAIN ** THE YOUNGSTOWN HOSPITAL ASSOCIATION PAGE 2

NORTH UNIT PATIENT CUMULATIVE REPORT DATE 08/24/77 TIME 08.49

PHYLLIS 8666016 AGE 51 F ADM DATE 07/29/77 PROV DIAG-HEPATITIS

REQUESTED LAB DATA *-HIGH #-LOW

WARNOCK R G MD

HONG J MD

		FRIDAY 07/29/77	SATURDAY 07/30/77	MONDAY 08/01/77	TUESDAY 08/02/77	WEDNESDAY 08/03/77	THURSDAY 08/04/77	FRIDAY 08/05/77
HOSPITAL DAY		1	2	4	5	6	7	8
COAGULATION								
PTT	0-45 SEC	80 *						
PRO TIME	70-100%N	0						
FIBRINOGEN	200-400 MGM%	93 #						
PLATELETS	150-350 TH/CMM	140 #						
FACTOR II	100%	10						
FACTOR VII	100%	10						
FACTOR X	100%	10						
THROMBIN TIME	9-11 SEC	35 *						
FIBRIN SPLIT PRODUCTS		NEG						
FACTOR VIII ASSAY	80-100%			100				...
CARDIOVASCULAR-HYPERTENSION								
SGOT	5-40 MU/ML		1200 *		1640 *			1280 *
CPK	25-145 MU/ML		25		29			33
LDH	100-225 MU/ML		380 *		446 *			399 *
CHOLESTEROL	150-250 MGM%		155					
TRIGLYCERIDE	50-150 MGM%		115					
LIVER FUNCTION								
BILIRUBIN TOT	0.1-1.5 MGM%	10.2 *	13.0 *		28.0 *			27.5 *
BILIRUBIN CONJUG	0-0.4 MGM%		5.8 *		16.0 *			12.0 *
ALK PHOS	30-110 MU/ML	141 *	143 *		165 *			201 *
SGPT	5-45 MU/ML		1340 *		1800 *			1420 *
PROTEIN TOT PLAS	6.2-8.4 G%		6.3		7.0			6.8
HEPATITIS B SURFACE ANTIGEN		NEGATIVE						
AMMONIA	0-70 MCG%	54 9AM						
PROTEIN ELECTROPHORESIS SER								
TOTAL PROTEIN	6-8 G%	4.9 #						
ALBUMIN	3.5-4.7 G%	2.6 #						
GLOBULIN	2.5-3.3 G%	2.3 #						
A/G RATIO	1.0-1.5	1.2						
ALPHA 1	0.2-0.4 G%	0.1 #						
ALPHA 2	0.6-1.2 G%	0.3 #						
BETA	0.6-1.2 G%	0.5 #						
GAMMA	0.7-1.4 G%	0.4 #						
PATTERN		POLYCLONAL						

PATHOLOGIST CONSULTATION 08/01 NOTE PAST HISTORY 5-5-65 ADENOCARCINOMA, PAPILLARY, THYROID--RAPPOPORT/EP

ANATOMIC PATHOLOGY

GROSS AND MICRO B 77-09071-N

PATHOLOGY DIAGNOSIS 08/20 SKELETAL MUSCLE AND FAT, NO LIVER TISSUE IDENTIFIED--NATIVIDAD/EP

INTERPRETATIONS

PATHOLOGIST CONSULTATION 08/24 HYSTERECTOMY FIVE WEEKS PREVIOUSLY. NO BLOOD TRANSFUSIONS. THREE DAYS PRIOR TO ADMISSION HAD NAUSEA, WEAKNESS, JAUNDICE. PHYSICAL EXAMINATION, X-RAYS OF CHEST AND OF ABDOMEN WERE NEGATIVE. JAUNDICE INCREASED IN SEVERITY ASSOCIATED WITH MARKED ABNORMALITIES OF ALL RELEVANT LIVER ENZYMES ALTHOUGH SGOT AND LDH LISTED UNDER CARDIAC PROFILE. EEG DISPLAYED METABOLIC ENCEPHALOPATHY. PROGRESSIVE COAGULATION DEFECT WITH ELEVATED PTT AND THROMBIN TIME WITH DIMINISHED FIBRINOGEN. INCREASING RENAL INSUFFICIENCY. HYPOPROTEINEMIA WITH POLYCLONAL GAMMOPATHY. ETIOLOGY OF HEPATITIS UNDETERMINED. NOTE HISTORY OF PAPILLARY CARCINOMA OF THYROID WITH POSSIBILITY OF METASTATIC DISEASE TO LIVER. LIVER BIOPSY UNSUCCESSFUL. FINAL DIAGNOSIS - SEVERE HEPATOCELLULAR NECROSIS WITH ASSOCIATED HEPATO-RENAL SYNDROME, METABOLIC ENCEPHALOPATHY AND COAGULOPATHY--A. E. RAPPOPORT, M.D./PW

Figure 6. A comprehensive example of a cumulative report indicating various PALM profiles. Notice the horizontal chronological arrangement of results for each test and the vertical arrangement of PALM Profile tests. Note the entry from the tumor registry indicating a previous diagnosis of thyroid carcinoma. The accession number of the current surgical specimen of an unsuccessful liver biopsy is stated. The interpretation by the pathologist represents an abstract of clinical findings integrated with the laboratory data.

Cumulative Patient Summary Report

This report is generated at 6.00 p.m. for each patient each day that a test has been performed. Figure 6 displays a typical report. It states the date of admission, the patient's name, age and sex and his provisional admitting diagnosis. If tests are not performed, a report is not printed that day. It permits chronological comparison of individual test results for seven test days in a horizontal arrangement.

Tests are listed vertically according to PALM. This permits more efficient integration and interpretation of all results in terms of the basic disease; it eliminates the possibility of overlooking the vital data (especially if abnormal), decreases unnecessary testing and simplifies the differential diagnosis.

Normal ranges are stated in the test description but the results are adjusted for age and sex. Their values are signified by special symbols if they fall outside of these ranges, i.e. an asterisk (*) indicates high abnormal, a double dagger sign (‡) indicates low abnormal.

In addition to presenting clinical pathology data, the cumulative report can also include unstructured, free text such as: the accession number and diagnosis of surgical pathology examinations; historical data from the tumor registry of previous history of malignancy; consultations by pathologists interpreting and evaluating the significance of the laboratory data in the light of clinical information as well as of X-ray, EEG or EKG findings which may be correlated with the results of cardiac enzymes or other studies. Cumulative reports are available on Visual Display Units as well as being printed. Such instant access to patients clinical data held in the PMF is useful to, say, a pathologist for evaluating when he is doing microscopic examinations of tissues or smears.

The Cumulative Summary Reports and consolidated lists of all abnormal values from each operating division are reviewed by the pathologist in charge of that division to check the data in the light of clinical information and to assist in the interpretation of results in particuarly complex or difficult cases. In such instances, the pathologist (consultant) furnishes his residents (registrars) with copies of the printouts with the request that medical rounds be made to the patients' station to obtain additional pertinent clinical information properly to correlate these data. Residents carry out such bedside analyses with the pathologists to determine whether or not additional studies or consultations are indicated. The pathologist will then dictate his opinions which are filed in the patient's record. These are transcribed by a secretary using a communicating magnetic card selectric typewriter (CMC/ST) connected by data set to the 370.

Presentation of data in such a logical and organised fashion permits the physician to focus rapidly and accurately on the salient information required for intelligent management of the case. It should be obvious that by this method of information interchange, coordination of effort between clinic and laboratory has been brought to a high level of efficiency and medical usefulness. The cooperation leads to closer, more intelligent use of the clinical laboratory by the attending physician who also obtains a better insight into the laboratory's potential for enhancing medical care by this joint venture.

Through the use of specialised software packages (Mark IV, Informatics, Inc.) and a locally developed program, the pathologist may request and receive printouts of special studies he considers necessary properly to control the performance of his laboratory for teaching or for research. Over 16,000 accessed cases of malignancy are available for prompt retrieval and listing by any of many diverse parameters, i.e. name, age, sex, clinical diagnosis, surgical pathology, diagnosis, topography, etiology, surgical procedure, chemotherapy treatment, recurrences, prognosis, laboratory tests, etc. Mark IV allows a large host of other printouts to be generated including annual followup requests to MDs, analyses of survival, types of disease or treatment, etc.

ANATOMIC PATHOLOGY

Although emphasis in the foregoing discussion has been placed on the impact of EDP in clinical pathology, anatomic pathology has not been neglected. As Figure 6 displays, the surgical pathology accession number and diagnosis are also conveyed to the cumulative summary report along with chemistry, hematology and other information allowing integration of those data with the anatomic findings. In addition, all hospital admissions are screened by the tumor registry staff for historical data of previously diagnosed malignancies or 'premalignant' conditions such as adenomas of liver (with 'the pill'). The responsible senior pathologist also receives a printout of consultations emanating from his division and, as a result, has opportunity to consider not only the accuracy of his diagnosis but its significance in terms of the total clinical aspects of the case.

Diagnosis Coding

Significant improvements have been achieved in the filing, storage and retrieval functions of anatomic and clinical diagnoses through the introduction of word processing techniques, rather than by the conventional use of numeric codes such as SNOP, ICDA, etc.

Immediately after the final diagnosis is dictated by the pathologist, transcription of the report is performed by the secretaries using the CMC/ST to prepare the complete document and to transmit by data sets, patient information and pathology diagnoses to each patient's file in the host computer. These entries are now available for retrieval according to case, name, diagnosis or other parameters, including age, sex, recurrence, type of operation, surgeon, consultant, etc. Mark IV programs have allowed us to perform advanced, rapid and efficient search techniques of all data available in the host.

New diagnoses or synonyms, accumulating from the daily operations of the anatomic pathology division, are filed in free text. These are stored archivally on magnetic tape which grows with the receipt of unique terms as they occur so that it represents a living, expanding and flexible data base.

The entire file can be searched rapidly for all cases bearing specific words designating anatomic site, organ, diagnosis, specific etiologic agents, drugs, organisms, etc., and listings may be generated for any combination of morphology, topography, etiology and clinical diagnoses. The size of each group can be contracted to refer to a single, specific parameter or expanded to include a large host of associated features under study. Synonym links, i.e. enlarged thyroid, adenoma thyroid or goiter, as previously specified by the pathologists, permits automatic search by name for other possibilities.

AUG 23, 1977

THE YOUNGSTOWN HOSPITAL ASSOCIATION
DEPARTMENT OF PATHOLOGY AND LABORATORY MEDICINE
STUDY OF REQUESTED SURGICAL DIAGNOSES
-- THYROID -- GOITER --

PAGE 4

PATIENT NAME	SURGICAL ACCESSION NUMBER	MEDICAL RECORD NUMBER	DATE OF SURGERY	DATE OF BIRTH	AGE AT SURGERY	DIAGNOSIS
DONALDSON LYNN	77-07075	7700636	06/29/77	08/27/44	32	06/30 CARCINOMA, FOLLICULAR, MULTIFOCAL, RIGHT AND LEFT LOBES, THYROID GLAND --LYMPH NODES, REGIONAL X2, METASTATIC CARCINOMA PRESENT--SABADO/JP
DUNCAN GEORGE	77-06835	8643636	06/24/77	07/02/47	29	06/24 FOLLICULAR ADENOMA, DEGENERATIVE OF THYROID --THYROIDITIS, CHRONIC, LYMPHOCYTIC, FOCAL, MILD--GILLIS/PC
DUNKLE DAVID A	77-06369	3637353	06/14/77	05/13/41	36	06/14 THYROID, SUBTOTAL THYROIDECTOMY SPECIMEN--THYROID ADENOMAS, SEVERAL--DEPPISCH/PW
EVANS ANGELA	77-02850	8581479	03/17/77	03/01/47	30	03/17 ADENOMAS, FOLLICULAR FETAL TYPE, LEFT LOBE OF THYROID --GOITER, ADENOMATOUS WITH FOCAL HYPERPLASIA, LEFT LOBE OF THYROID --LYMPH NODE NO PATHOLOGIC DIAGNOSIS, PART B --CICATRIX, OF SKIN--NATIVIDAD/PW
FRAZIER JOHN	77-02295	8573301	03/03/77	04/24/47	29	03/03 ADENOMA, FOLLICULAR, LEFT LOBE, OF THYROID --THYROID GLAND, PORTION OF, RIGHT, NO PATHOLOGIC DIAGNOSIS--GILLIS/ALR/PW
FREEMAN MARTHA	77-06973	7700105	06/28/77	03/27/25	52	06/28 ADENOMA, FOLLICULAR, RIGHT LOBE OF THYROID--GILLIS/PW
GAVIN OPAL	77-04149	8602603	04/20/77	03/02/36	41	04/20 ADENOMATOUS GOITER, SUBTOTAL THYROIDECTOMY--NATIVIDAD/AME
GIBBS ETHEL L	77-04466	7671687	04/27/77	03/06/54	23	04/27 THYROID, SUBTOTAL THYROIDECTOMY SPECIMEN--HYPERPLASIA, SEVERE--DEPPISCH/AME
GRIFFITH WILLIAM	77-02898	8582114	03/17/77	08/21/59	17	03/17 CARCINOMA, FOLLICULAR, LEFT LOBE OF THYROID --CARCINOMA, FOLLICULAR, METASTATIC TO 4 OF 5 LYMPH NODES AND SOFT TISSUE NODULES, MULTIPLE, PARATRACHEAL, AND 2 OF 2 LYMPH NODES, MEDIASTINAL --THYMUS, 2 PORTIONS OF, NO PATHOLOGIC DIAGNOSIS --PARATHYROID GLAND, X2, NO PATHOLOGIC DIAGNOSIS --PYRAMID AND RIGHT LOBE OF THYROID, NO PATHOLOGIC DIAGNOSIS--NATIVIDAD/JP
GRISDALE EDWARD	77-07573	8655065	07/13/77	10/02/06	70	07/13 THYROID GLAND, RIGHT LOBE--ADENOMATOUS GOITER--DEPPISCH/PW
HADDOX LEONA	77-04551	8607737	04/28/77	11/18/20	56	04/29 GOITER, ADENOMATOUS WITH CALCIFICATION, HEMORRHAGE AND CYSTIC DEGENERATION, SUBSTERNAL, PARAMEDIAL AND RIGHT LOBE OF THYROID GLAND--SABADO/PB

Figure 7. This is a partial listing by patient name of cases of thyroid disease (carcinoma, adenoma, hyperplasia) filed in the surgical pathology department.

Advantages of this free, unstructured, non-coded filing and retrieving system are numerous. The tedious, error-prone, manual coding and indexing by numbers obtained from looking up various indices is entirely eliminated. Teaching, research and review of data is facilitated by the ability to retrieve series of cases by means of specified parameters. There are no constraints on the type or number of characteristics which can be stored and integrated with each other.

Figure 7 indicates a partial listing of diseases of the thyroid (goiter) abstracted by this method from the files. This example has been printed in patient alphabetic order but could also be displayed by pathology accession number, date, patient age, or type of lesion at the option of the requesting physician.

Laboratory Administration

Not only are there pressing medical needs for processing professional information, but laboratory management also demands much of EDP to establish budgets, control staffing, analyse equipment needs, assess use of supplies and determine efficiency of the total operation. High on the list of priorities is the need to eliminate patient, specimen or result identification errors and to guarantee precision, accuracy and maximum quality assurance of all steps along the test path.

By use of the computer and appropriate software, all activities may be logged-in and analysed to measure the elapsed time between each step. Controls can be designed to insure that delay is eliminated, that the integrity of the specimen and test be maintained and that the identity of the worker be recorded.

In this regard, a major managerial achievement of computerised laboratories has been significant improvements in the patient and specimen identification systems. The source of many errors in the laboratory, misidentification is disappearing rapidly as a result of machine readable identification and audio-verification systems.

PATIENT IDENTIFICATION

On being admitted to the hospital, each patient receives a plastic wristband with a transparent pouch possessing a card bearing his name and his physician's name, his age, sex, room number and hospital number (see top of Figure 3). Fixed to that band is a row of small detachable pressure labels imprinted with the patient's hospital number. These labels are removed and fastened to all containers of specimens derived from that patient including urine, sputum, feces, spinal fluid or gastric secretions when sent with an appropriate requisition to the laboratory. The phlebotomist also removes these labels and attaches them to the tubes bearing the human and machine readable stub card described above. For hematology (CBC), extra labels are removed for attachment, one to the stained smear of the differential count and a second to a port-A-punch card used to enter the results of that examination into the computer.

Currently, we are altering the print style of those labels to conform to Font A for OCR, Optical Character Reading, so that these numbers can be read by Wand, as recommended by Professor Dudeck and others in this book.

These safeguards, plus audible verification of test and patient names and special conditions through DIVOTS lead to markedly enhanced security of identification.

CONCLUSION

The foregoing description of the current state of the art of EDP in the department of pathology and laboratory medicine of the Youngstown Hospital Association has stressed the development of electronic data acquisition, communication and information systems as tools of physicians and administrators to enhance medical benefits and cost effectiveness.

Through the employment of the telephone audio response system (DIVOTS), automated test performance and computerised data reduction, all steps between the physician's order and the submission of a test result are rapidly executed with elimination of delay and error, thus permitting early physician decision-making in the care of his patient, leading to reduction of length of hospital stay and significant lowering of cost.

The intelligent presentation of integrated, clinically relevant test results by voice or display by printed page–problem assisted laboratory medicine (PALM), increases the efficiency of the physician, reduces errors of omission by eliminating failure to order or to note the results and permits him to organise his medical management strategy in an acceptable fashion.

Retrieval of vital information for intralaboratory control, teaching and research is simplified and expedited.

This description also affords insight into a building block method of development by which computer coupled automation may be introduced successfully into a working laboratory if prudence and patient and backup procedures are available. These are the dykes which hold back the flood of data and sluice the information into proper and manageable channels.

ACKNOWLEDGEMENTS

This project was supported by HEW Grant Number R18 HS 00060 from the National Center for Health Services Research, HRA.

FOR FURTHER READING

Rappoport, AE, Constandse, WJ, Seligson, D and Greanias, EC (Ed) (1965) *Symposium on Computer Assisted Pathology,* Miami, Florida, College of American Pathologists, Chicago.

Rappoport, AE, Gennaro, WD and Constandse, WJ (1967) Cybernetics Enters The Hospital Laboratory, *Mod. Hosp. 108,* 107

Rappoport, AE, Gennaro, WD and Constandse, WJ (1968) Computer-Laboratory Link is Base of Hospital Information System, *Mod. Hosp. 110,* 94

Rappoport, AE and Rappoport, E (1970) Laboratory Automation, *Hospitals, 44,* 114.

Rappoport, AE (Ed) (1971) Computering Without Tears In *IV International Symposium on Quality Control,* Hans Huber, Berne, Page 114

Rappoport, AE (1973) Computers, Information Retrieval and Data Storage In *Lab. Med.* (Ed) GJ Race. Harper and Row, New York. Chapter 23, Page 1

Rappoport, AE and Gennaro, WD (1974) The Economics of Computer-Coupled Automation in the Clinical Chemistry Laboratory of The Youngstown Hospital Association, In *Computers in Biomedical Research,* (Ed) RW Stacy and BD Waxman. Academic Press, New York. Chapter 4, Page 215

Rappoport, AE, Gennaro, WD and Berquist, RE (1975) An On-Line Centralized Computer-Coupled Automated Laboratory Information System Using Touch-Tone Card Dialer Telephone and Audio-Response Technology for Test Order Entry and Result Retrieval, In *Proceedings, National Computer Conference, Anaheim, Calif.* AFIPS Press, Montvale, N.J. *44,* Page 757

Rappoport, AE, Gennaro, WD and Berquist, RE (1975) A Computerized, International, Audio-Response Telephone Communication System Serving a Centralized Computer-Coupled, Automated Clinical Laboratory. (presented at the 39th Annual Convention CSLT) *Canad. J. Med. Tech., 37,* 150

Rappoport, AE, Gennaro, WD and Berquist, RE (1975) The Communicating Magnetic Card as a Word Processor to Enter Pathology Narrative Reports and Clinical Laboratory Results into a Patient's Computer Record, *Lab. Med. 6,* 22

Rappoport, AE, Gennaro, WD and Berquist, RE (1975) A First Trans-Atlantic, Live Demonstration of a Computerized Audio-Response Communication System of an Automated Clinical Laboratory, In *Quality Control in Clinical Chemistry* (Transactions of the VIth International Symposium, Geneva, Switzerland, April 1975), (Ed) G Anido. Walter de Gruyter, West Berlin 30, Page 353

Rappoport, AE, Gennaro, WD and Berquist, RE (1975) A Live, Trans-Pacific Demonstration of an Audio-Response Communications System of an Automated Computerized Clinical Laboratory, In *Proceedings of MEDIS '75 International Symposium on Medical Information Systems* (Tokyo, Japan, October 1975) The Medical Information System Development Center, Kansai Institute of Information Systems, Page 354

Rappoport, AE, Gennaro, WD and Berquist, RE (1975) An Audio-Response Communications System for Order Entry and Result Retrieval in an Automated Computerized Clinical Laboratory, *The Effect of Environment on Cells and Tissues* (IX World Congress of Anatomic and Clinical Pathology, Sydney, Australia, October 1975) Excerpta Medica International Congress Series Amsterdam, No. 384, Page 223

Keller, H (1975) Der Sprechende Labor-Computer: Ein Neuer Weg der Informations-Ubermittlung Zwischen Arzt und Labor, *Med. Lab. 28,* 1

Lewis, HL (1975) 2001 Today: Computer That Talks Tells Doctors Laboratory Findings, *Mod. Healthcare, 3,* 16

Rappoport, AE (1973) Computer, Information Retrieval and Data Storage, In *Laboratory Medicine,* (Ed) GJ Race. Harper and Row, New York. Chapter 23, Page 1 (Revised and updated in Addendum, 2nd Edition, 1976)

Rappoport, AE, Berquist, RE and Gennaro, WD (1976) Benefits of an Automated Computerized, Centralized Clinical Laboratory with a Telephone Audio-Response Communication System for Order Entry and Result Retrieval, In *Common Goals: Working Together,* Proceedings of the 6th Annual Conference of the Society for Computer Medicine (Conference Edition), Boston, Mass. *5.1,* Page 1

Rappoport, AE, Gennaro, WD and Berquist, RE (1977) A Telephone Audio-Response Communication System Serving a Centralized Automated, Computerized Clinical Laboratory, *MEDCOMP '77,* International Congress on Computing in Medicine, Berlin. Page 209

Rappoport, AE, Gennaro, WD and Berquist, RE (1976) An On-Line Centralized, Computer-Coupled Automated Laboratory Information System Using Touch-Tone Card Dialer Telephone and Audio-Response Technology for Test Order Entry and Result Retrieval, *Lab, III,* (3) 284

Rappoport, AE (1976) Automated Laboratory Testing System Reports Results Verbally, *Computer Design,* April, 1976

Rappoport, AE, Gennaro, WD and Berquist, RE (1976) A Centralized Automated and Computerized Clinical Laboratory Served by a Touch Tone Audio Response Telephone System for Test Order Entry and Result Retrieval, *Quality Control in Laboratory Medicine,* JL Giegel and JB Henry (1st Interamerican Symposium on Quality Control in Lab. Med., Key Biscayne, Florida, April 1976)

Rappoport, AE, Gennaro, WD and Berquist, RE (1976) Benefits of an Automated, Computerized, Centralized Clinical Laboratory With a Telephone Audio Response Communication System for Order Entry and Result Retrieval, *Japanese Jour. of Clin. Lab. Automation* (JJCLA), 1

GETTING THE BEST VALUE FROM A LABORATORY

Professor Hannes Buttner Dr Med, Dr rev nat, Dipl chem
Professor of Clinical Chemistry, Institute of Clinical Chemistry, Medical High School, Hannover, Germany

INTRODUCTION

The phrase 'best value from a laboratory' may be interpreted in widely differing ways. To the clinician it may mean getting data that are relevant to the concrete diagnostic question; the clinical chemist will think in terms of the particular reliability of the results; and the health service administrator's interest will attach to cost/benefit figures. In relation to the target of clinical laboratory investigations as set out in Figure 1, these three different but equally valid aspects may be succinctly called 'reliability', 'information relevant to diagnosis', and 'economics'.

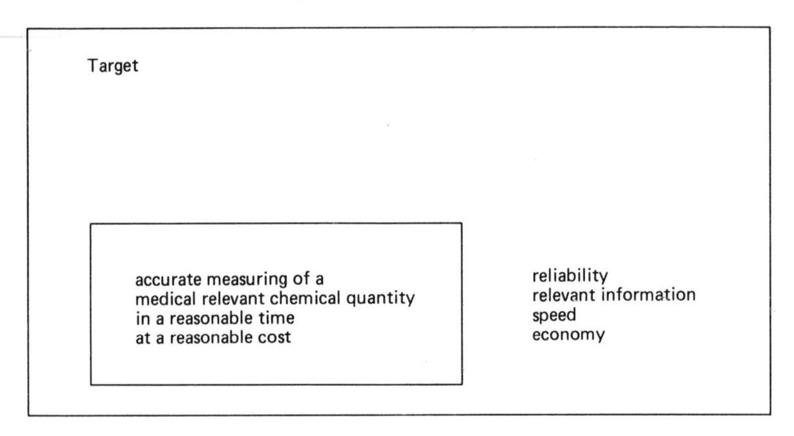

Figure 1. Target of clinical chemical analyses

During the last 15 years, a considerable amount of work has been done in clinical chemistry to ensure and improve analytical reliability, particularly also with a view to bringing in Electronic Data Processing (EDP). Efforts to obtain relevant diagnostic information at reasonable costs have only just started, however. The problems involved are those of correct interpretation, and the selection of tests which are suitable also from the cost/benefit point of view. Their solution will be possible only with the aid of EDP.

In this paper I intend to discuss certain basic aspects in the optimisation of information derived from diagnostic tests, and to present a number of possible solutions which apply also to routine operation.

Evaluation	Longitudinal	Transversal
individual result compared with	earlier data from same individual	reference population
decision categories	changed/not changed	diseased/not diseased

Figure 2. Data transformation or interpretation of analytical results

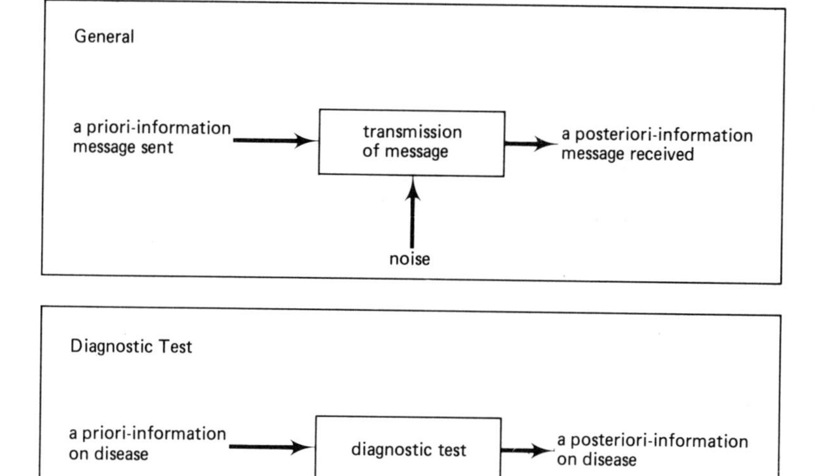

Figure 3. Diagnostic test as an information process

THE INFORMATION CONTENT OF LABORATORY DATA

Laboratory data such as 'serum potassium concentration 4.0 mmol/l' are of no value in themselves. A transformation process known as 'interpretation' is required to convert them into clinically relevant information (1, 2, 3). This transformation normally takes the form of either longitudinal or transversal evaluation (Figure 2).

To optimise the transformation process, the information which has been obtained needs to be quantified. This may be done by using the methods of information theory (4, 5). The basic principle and its application to diagnostic tests are shown in diagrammatic form in Figure 2 (6, 7, 8). The prior information available on the presence of disease is increased by the test, with test errors taken into account as 'noise'. As an example, a more detailed description is given below of transversal evaluation. For purposes of simplification, the test result shall be assumed to be in the form of a binary statement ('diseased'/'not diseased') obtained with the aid of a decision criterion, rather than a continuous variable. A typical example applicable to many clinical laboratory tests is shown in Figure 4. In the terminology of information theory, it may be said that the signal 'disease' (D) is to be recognised in the presence of noise ('non-disease' (d) in this

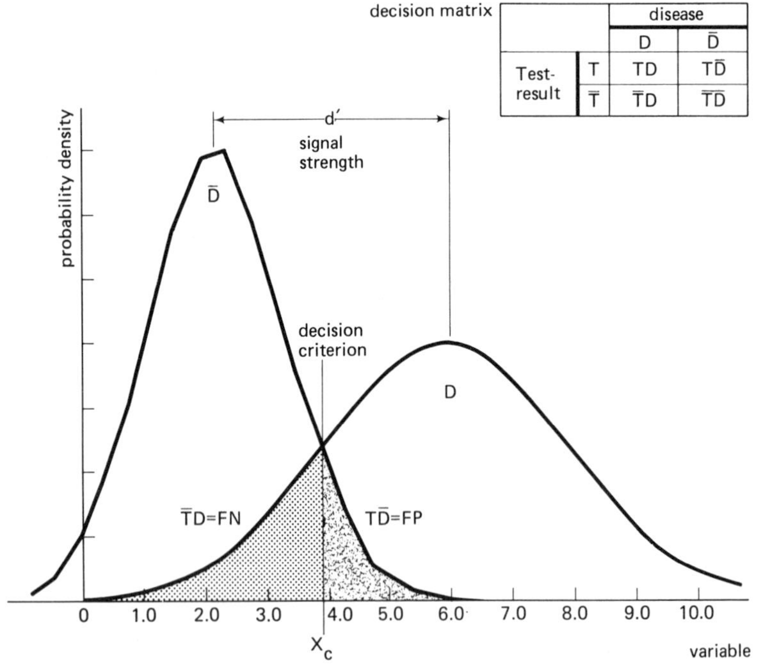

Figure 4. Use of a decision criterion Xc for interpretation of quantitative analytical results. Overlapping of distribution curves results in false positive (FP) and false negative (FN) decisions. (see reference 8) D = diseased, d = not diseased, T = positive test result, t = negative test result

case). The results obtained will be better the smaller the overlap between populations D and d (i.e. the greater the signal strength d). A quantitative measure of information used in information theory is entropy (Figure 5) which is calculated from the prior and posterior probabilities (8). In the case of a diagnostic test (Figure 6) information gain depends on the one hand on prevalence (P(D) = prior probability). If this is 1, i.e. certain, additional information through the test is not possible. On the other hand information gain is determined by the true positive rate (P(T/D)) and the

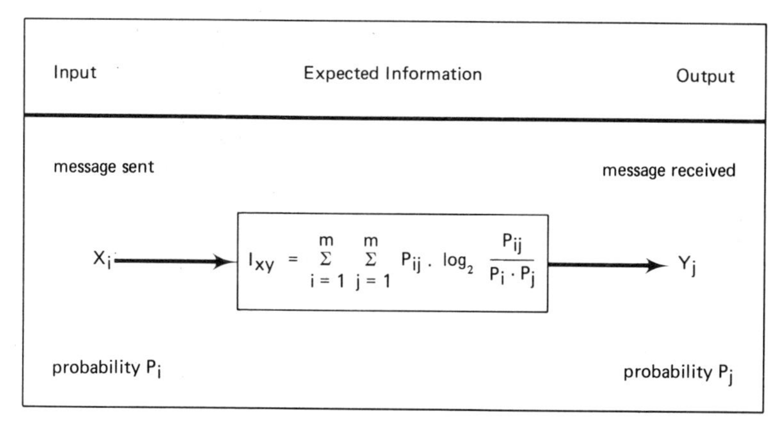

Figure 5. Information content of an indirect message (for details see reference 8)

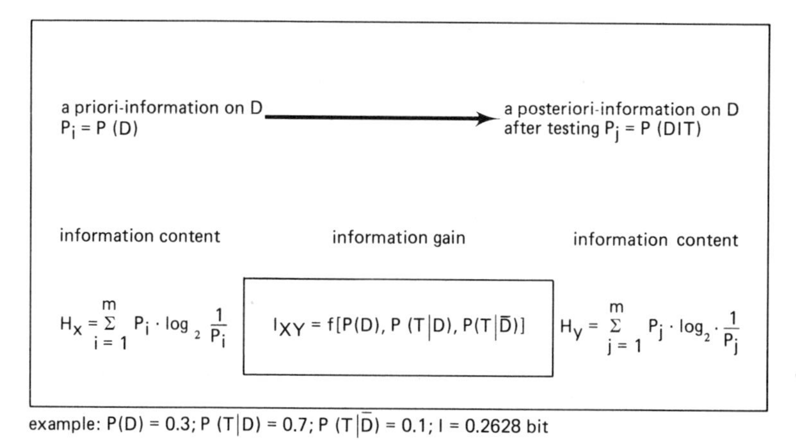

example: $P(D) = 0.3$; $P(T|D) = 0.7$; $P(T|\bar{D}) = 0.1$; $I = 0.2628$ bit

Figure 6. Information content of a diagnostic test. Probability notations (8): P(D) = Probability of disease = prevalence; P(T/D) = conditional probability of positive test result T, given D = sensitivity of test, P(T/d) = probability of false positive results (= 1-specificity). Algorithm and example see reference 8.

false positive rate (P(T/d) = noise). The information gain I of a diagnostic test serves as a useful quantitative assessment of the properties of a diagnostic test, particularly in connection with the optimisation of tests which is discussed below (9).

OPTIMISATION OF INFORMATION GAIN

The provision of relevant information on reasonable costs has been stated as one of the targets of clinical laboratory tests, This calls for the optimisation of information gain in the procedure of interpretation. For this important requirement we have to rely particularly on EDP.

In the case of transversal evaluation with binary statement, which has already served as an example, optimisation involves optimal definition of the decision criterion (Xc in Figure 4), taking into account prevalence, methodological errors, costs and benefits.

As a first approximation, let us consider the effect of variable prevalences on the decision criterion in the case of normal distribution (Figure 7). Misclassification will obviously be minimised by placing the decision criterion at the point of intersection for the distribution curves. The position of Xc is then easily calculated for various prevalences P(D).

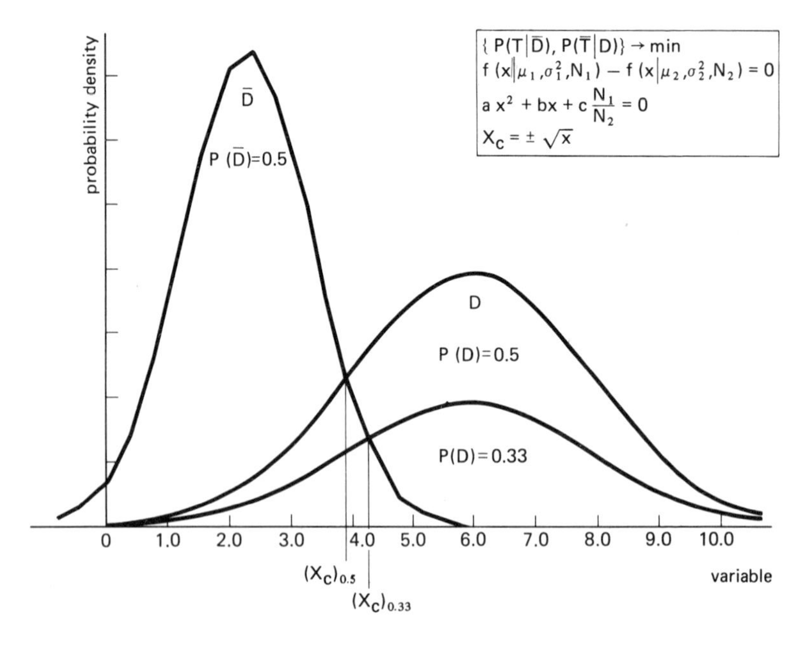

Figure 7. Decision criterion Xc for variable prevalences of disease. The populations d and D are considered to be gaussian. N_1, N_2 population sizes

A more comprehensive and realistic approach to optimisation must include consideration of costs and benefits (7, 8, 10). This raises a number of problems. The benefits of medical measures are difficult to quantify, being based on both objective and subjective assessment. The calculation of costs is equally complex, as it involves indirect as well as direct costs.

	False Positive	False Negative
Direct	cost of the test (incl. repetition), following diagnostic, unneccessary therapy	cost of test repetition and following diagnostic measures
Indirect	risk of the test, consequences of false diagnosis and unnecessary therapy, cost of apparently diseased	consequences of non-detection of a disease (especially preventive medicine) and non-treatment, cost of premature death or permanent disability

Figure 8. Cost or negative utilities of false positive and false negative test results

Payoff Matrix (Utilities)

		disease	
		D	$\bar{D}$
test	T	U_{TD}	$-U_{T\bar{D}}$
	$\bar{T}$	$-U_{\bar{T}D}$	$U_{\bar{T}\bar{D}}$

Expected Utility

$$E(U) = U_{TD} \cdot P(T|D) \cdot P(D) + U_{\bar{T}\bar{D}} \cdot P(\bar{T}|\bar{D}) \cdot P(\bar{D}) + U_{T\bar{D}} \cdot P(T|\bar{D}) \cdot P(\bar{D}) + U_{\bar{T}D} \cdot P(\bar{T}|D) \cdot P(D)$$

Figure 9. Expected utility of diagnostic testing utilities U of the possible outcome are given in a payoff matrix. The expected utility is calculated using sensitivity, specificity, and error rates of the test

Direct and indirect costs will tend to be high if false positives or false negatives are obtained. Figure 8 shows some of these costs. For costs and benefits to be included in the optimisation procedure, it is useful to produce a 'pay-off matrix' for all possible tests results. This is done uniformly in terms of (positive or negative) utility (Figure 9) (11). Such utilities can be determined only on the basis of extensive data material, and at the present time this is not available. The literature contains only occasional references (8, 12). To demonstrate the effect of the error rate of a test and of the prevalence of disease on the expected utility of a test, Figure 10 shows the calculation of a theoretical example. The utilities

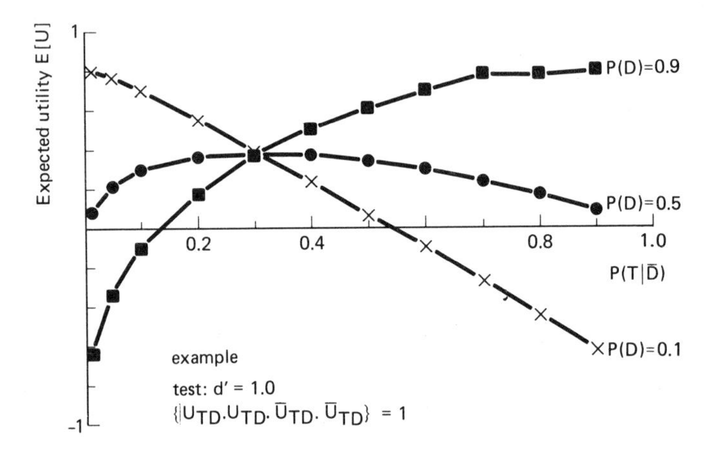

Figure 10. Influence of false positive rate and prevalence on expected utility. The example is calculated using equation from Figure 9 for a test with signal strength d' = 1 and gaussian distributions. All utilities are set to /1/

for the pay-off matrix were all put at 1 for this purpose. It will be noted that for maximum expected utility the test system will have to be adjusted to the conditions pertaining (position of decision criterion Xc). Appropriate methods will therefore have to be found to calculate the optimal position of the decision criterion Xc. A suitable method has been developed in the field of signal detection theory (13, 14). This involves producing a receiver operating characteristic curve (ROC curve) for the test system, to show the true positive rate of the test versus the false positive rate (Figure 11) (15). The optimum for each individual test system is represented by a point on this curve, and the information gain may then be calculated as described (Figures 5, 6). The optimisation of a test for minimal errors may be calculated with the aid of utilities (Figure 11) (13). This technique was

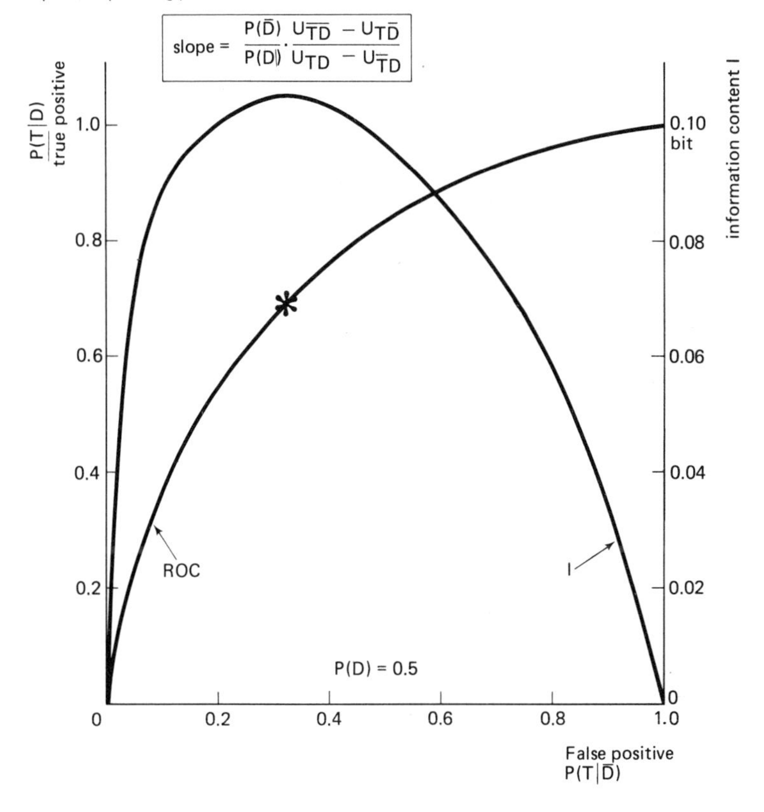

Figure 11. Test optimisation by means of a receiver operating characteristic (ROC) curve. The example is calculated for a test with d' = 1 and gaussian distributions. Information content I is calculated according to reference 8. Optimal operating point (13, 14) is marked by an asterisk

in fact used for the example shown in Figures 4 and 7. Figure 12 gives the results in a graphic representation designed to meet practical requirements. A major conclusion to be drawn from the optimisation procedure just described, is that the current practice of transversal evaluation of laboratory data is not adequate. The decision criterion is usually set at the upper or lower limit of the 95 per cent reference range–provided correctly defined

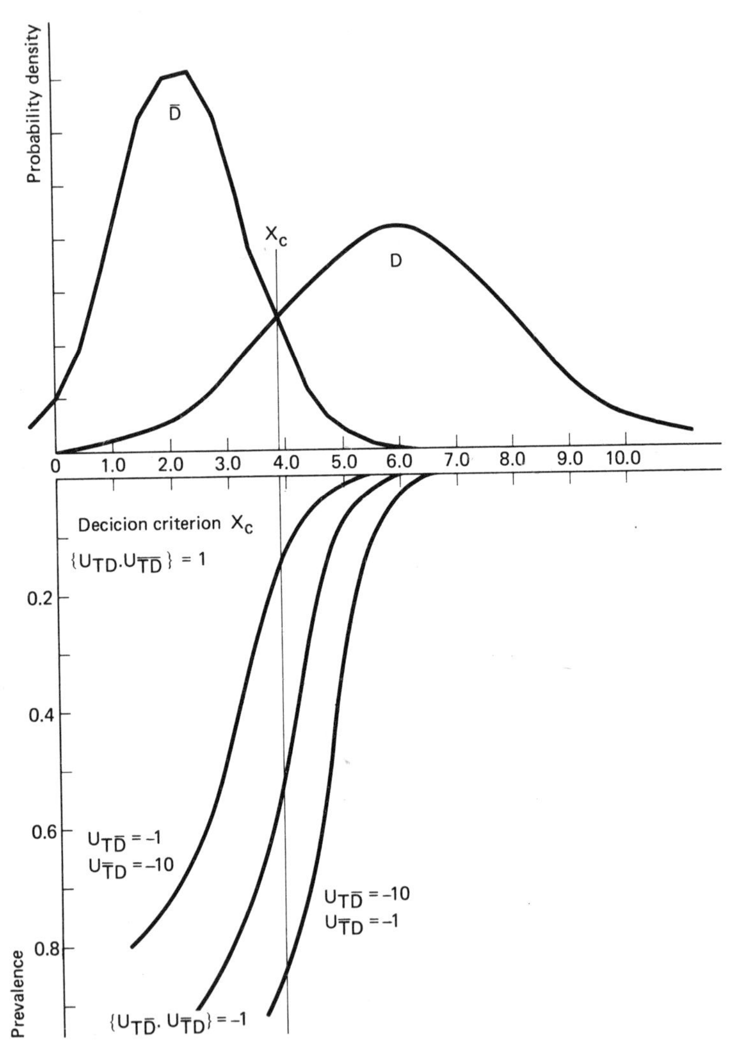

Figure 12. Optimal decision criterion of a test as a function of prevalence and utilities (test with d' = 1 and gaussian distribution). Calculation of optimal decision criterion according to Figure 7 introducing utilities

reference ranges have been used at all. A variable decision criterion adapted to the diagnostic situation would obviously be feasible only with computer-aided calculation and evaluation of analytical results. This represents an important future application of EDP in the interpretation of laboratory data. For routine purposes, consideration should also be given to the variable effect on the decision criterion of analytical precision and accuracy (16, 17, 18). For the sake of simplicity, this problem complex has been omitted from the present discussion.

OPTIMISATION OF MULTIPLE ANALYSIS

So far we have been concerned with the optimisation of single test interpretations. Theoretical considerations suggest that the diagnostic information value will increase if a number of suitable tests are used in

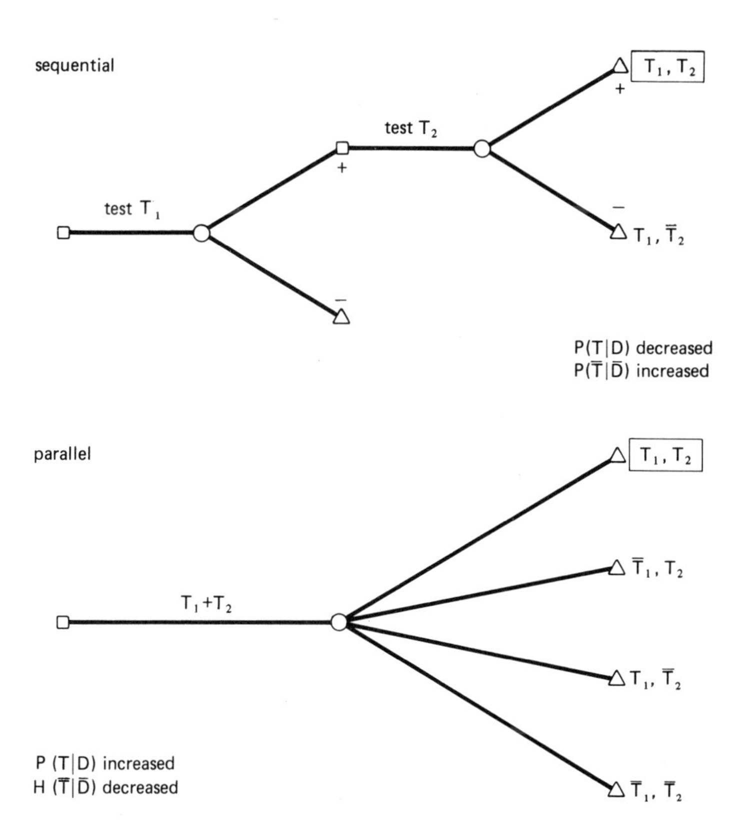

Figure 13. Sequential and parallel testing

combination. This should not, however, be done indiscriminately—e.g. by using a 'battery of tests'—as is so commonly the case. Various authors have discussed the problems arising with this practice (e.g. the high incidence of false positive) (8, 19, 20). What is required are multivariate techniques of statistical analysis, and occasional use is already made of these in routine diagnostic procedures. A detailed discussion of these techniques forms another part of this book, and we shall therefore limit ourselves to the problem of optimal test strategy with multiple analysis, using methods employed in decision theory. These methods permit the cost and benefit figures for different measures to be taken into consideration.

For the diagnostic utilisation of multple tests, considerable practical importance attaches to the question as to whether the tests are to be simultaneous (i.e. parallel) or sequential (Figure 13). By calculating specificity, sensitivity, and predictive value, it is easy to demonstrate that sequential testing results in a reduction in sensitivity but an increase in specificity (8, 19). With parallel testing, the effect depends on the defined decision matrix (one or both tests positive rates as positive).

A decision as to which strategy should be chosen must, in addition to the properties of the tests in question, also take into account the difference in utility of the results. Such decision problems may be solved in quantitative terms with the aid of a decision tree. The principle is shown in Figure 14. By taking into account the branch probabilities, the given utilities of the different results may be used to calculate the

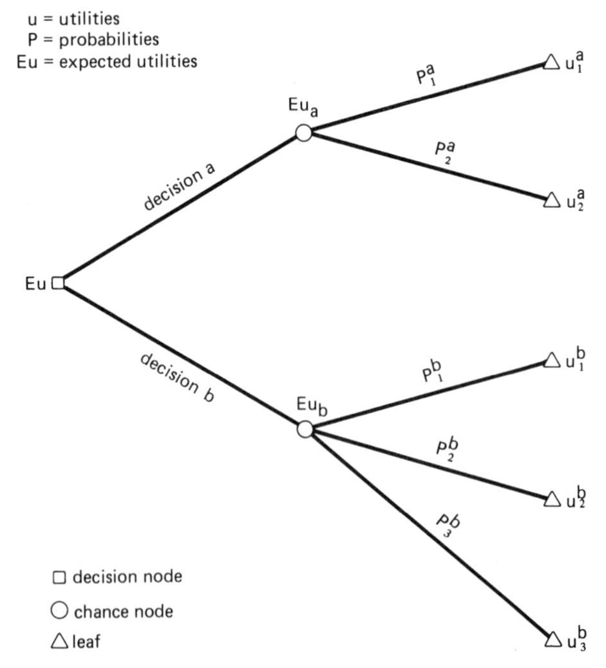

Figure 14. Principle of decision tree technique (for details see references 11, 22)

expected utility (11, 21, 22). For example the method may be applied to the following problem. In what sequence should creatine kinase (CK) and glutamate-oxaloacetate transaminase (GOT) determinations be done in a case where myocardial infarction is the tentative diagnosis–first CK and then GOT, or vice versa? The data for the test properties may be found in the literature (19). Figure 15 gives the full calculation. The utilities in this case are hypothetical, as the literature does not provide relevant detail. In the example, the decision arrived at is that the GOT test should come first, and, if the result is positive, the CK test second. This decision path gives the highest figure for expected utility.

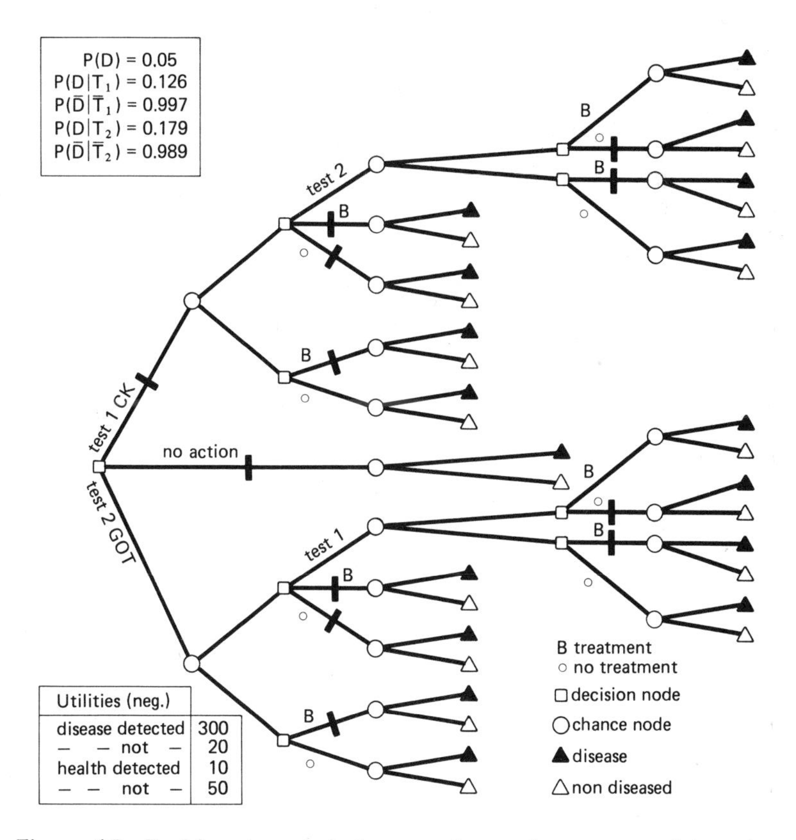

Utilities (neg.)	
disease detected	300
— — not —	20
health detected	10
— — not —	50

Figure 15. Decision tree technique used to solve a sequential testing problem: should CK or GOT testing be done firstly in cases of heart infarction

CONCLUSIONS

It is hoped that the examples in this paper have demonstrated that with the aid of modern mathematical techniques it is possible to optimise the information gain from clinical laboratory studies. A precondition for this is an adequate data base (not available at present) and the use of EDP. It has also been shown that it is possible with those techniques to take account of aspects relating to costs and benefit. Figure 16 is a diagrammatic representation of the pathway used to gain information, from the

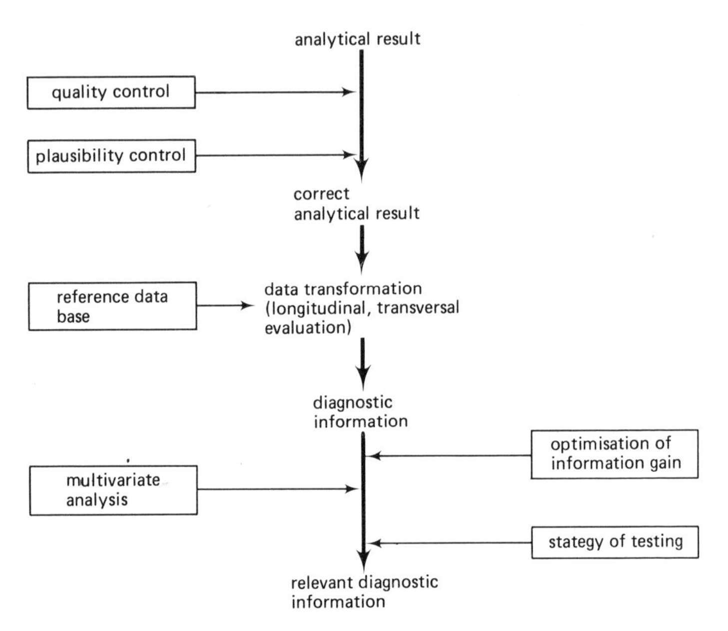

Figure 16. Relevant diagnostic information by use of EDP

results of a clinical laboratory analysis to the information relevant to diagnosis. It also shows at what points along this path EDP should be used. Clinical chemists have made great efforts in recent years to optimise quality control so as to ensure reliability. We should now pay increased attention to the improved interpretation of laboratory results.

REFERENCES

1. Buttner, H, Hansert, E and Stamm, D (1974) In *Methods of Enzymatic Analysis.* (Ed) Bergmeyer HU Verlag Chemie Weinheim. Academic Press Inc. New York and London. Vol 1, Page 318.
2. Young, D (1976) *Clin. Chem. 22,* 1555
3. Sunderman, FW jr (1975) *Clin. Chem. 21,* 1873
4. Shannon, CE and Weaver, W (1949) *The Mathematical Theory of Communication.* University of Illinois Press
5. Young, JF (1971) *Information Theory.* Butterworth and Co, London
6. Lusted, LB (1968) *Introduction to Medical Decision Making.* C.C. Thomas Publ., Springfield, Illinois

7. Barnoon, S and Wolfe, H (1972) *Measuring the Effectiveness of Medical Decisions.* C.C. Thomas Publ., Springfield, Illinois
8. Buttner, H (1977) *J. Clin. Chem. Clin. Biochem. 15,* 1
9. Metz, CH, Goodenough, DJ and Rossmann, K (1973) *Diagnostic Radiology, 109,* 297
10. Werner, M, Brooks, SH and Wette, R (1973) *Human Pathology, 4,* 17
11. Buhlmann, H, Loeffel, N and Nievergelt, E (1975) *Entscheidungs- und Spieltheorie.* Springer-Verlag, Berlin, Heidelberg, New York
12. Bay, KS, Flathman, D and Nestman, L (1976) *Amer. J. Publ. Health 66,* 145
13. Peterson, WW, Birdsall, TG and Fox, WC (1954) Transactions of the IRE Professional Group on Information Theory (PGIT) *4,* 171
14. Swets, JA (Ed) (1964) *Signal Detection and Recognition by Human Observers. Contemporary Readings.* John Wiley and Sons, New York, London, Sydney
15. Werner, M (1977) *Med. Welt 28,* 1254
16. Buttner, H (1973) In *Optimierung der Diagnostik* (Ed) Lang, H, Rick, W and Roka, L. Springer-Verlag, Berlin, Heidelberg, New York, Page 233
17. Holzel, W (1973) *Zbl. Pharm. 112,* 339
18. Martin, HF, Gudzinowicz, BJ and Fanger, H (1975) *Normal Values in Clinical Chemistry.* Marcel Dekker, Inc. New York
19. Galen, RS and Gambino, SR (1975) *Beyond Normality: The Predictive Value and Efficiency of Medical Diagnosis.* John Wiley and Sons, New York, London, Sydney, Toronto
20. Sunderman, FW (1970) *Amer. J. Clin. Pathol. 53,* 288
21. Chernoff, H and Moses, LE (1959) *Elementary Decision Theory.* John Wiley and Sons, New York, London
22. Lindley, DV (1971) *Making Decisions.* Wiley Interscience. London, New York, Sydney, Toronto

PLAUSIBILITY CONTROL IN CLINICAL CHEMISTRY

U Ludwig, Dr med,
Dedizinische Universitätsklinik,
Abteilung IV, D-7400 Tübingen, Germany

INTRODUCTION

Quality control with pooled serum is a well known method in clinical chemistry (1). It is applied to testing the accuracy and precision of analytical results. Its objective is to discover errors in method. These errors are caused by disorders in the instruments, faults in their operation or spoiled chemicals. On the other hand, it is possible that wrong or misleading laboratory results reach the doctor on the ward, although instruments, handling and chemicals are in order. The reasons are multiple: wrong identification of samples on the ward or in the laboratory, writing errors, influence of therapy and so on (2). All these errors are not detected by quality control and up to now have been found through intuitive plausibility control by the laboratory doctors (3). The number of our results has become too large for controlling intuitively and we had to search for a method to find these errors by computer.

COMPUTER AND PROGRAM

All laboratory data are manipulated and stored in a process computer, an IBM 1800. Patient number, date, hour and method number are stored together with each laboratory result. Programs for statistical research and for plausibility control are written in Fortran IV. The results of about 1000 patients are used to obtain the distribution and to compute the limits for plausibility control.

Abbreviations and methods of analysis

APh: alkaline phosphatase (4)
GOT: aspartate aminotransferase (5, 6)
GPT: alanine aminotransferase (7)
LAP: leucine arylamidase (8)
LDH: lactic dehydrogenase (9)
Ca: calcium (flame photometer)
Gluc: blood glucose (10)
Bil: total bilirubin (11)

COMPONENTS OF PLAUSIBILITY CONTROL

The laboratory results can be considered as an array of multidimensional data. The dimensions are:

1. patient,
2. time,
3. method,
4. result.

A given result can be compared with the results for other patients and the previous results for the same patient. It can also be considered in the light of other tests applied to the patient. From that the components of plausibility control are derived. We (12) have formulated them as:

1. extreme value control,
2. trend control,
3. pattern control.

Extreme value control is the comparison with results of other patients. Trend control is the examination of results for different times. Pattern control considers results from other methods.

COMPARISON BASIS

Each type of control is a comparison of the result with an expected or theoretical value. In quality control the theoretical values are the limits of previous measurements. Theoretical limits for plausibility control are: biological or statistical parameters, logical inconsistencies, comparison of two or more interdependent values, and variation between consecutive measurements.

LOGICAL LIMITS

Both logical and statistical limits are used in plausibility control. Some examples of logical limits are as follows:
The value for lactate dehydrogenase must be larger than the value for its isoenzyme-1 (α-hydroxy-butyrate dehydrogenase). The same goes for acid phosphatase and prostatic phosphatase, for total bilirubin and conjugated bilirubin. Other limits are given by sex. A woman has no prostatic phosphatase and for a man a positive pregnancy test should not be found.

STATISTICAL LIMITS

To define a statistical limit one would take a random sample of outcomes for a given test and from this determine the limit of say the 0.99 probability distribution. The use of 0.99 probability distribution is fixed as a practical limit. That means: If 1000 results are marked because they fall outside these limits, on the average 10 results are probably not wrong, but remarkable. One can compare these 0.99 probability limits with the 3 standard deviations limits of quality control, given a normal distribution of results for quality control.

EXTREME VALUE CONTROL

The unselected results for about 1000 patients are plotted as a histogram. Figure 1 shows such results for alkaline phosphatase and it can be plainly seen that in this case the distribution is not Gaussian. As an alternative to searching for a function that will describe the distribution in terms of parameters such as standard deviation we can adopt an empiric approach in finding distribution quantities. The data are searched for the limits of the middle 0.99 of the results. These limits are set as control limits for extreme values. Examples for extreme value control limits are given in Table 1.

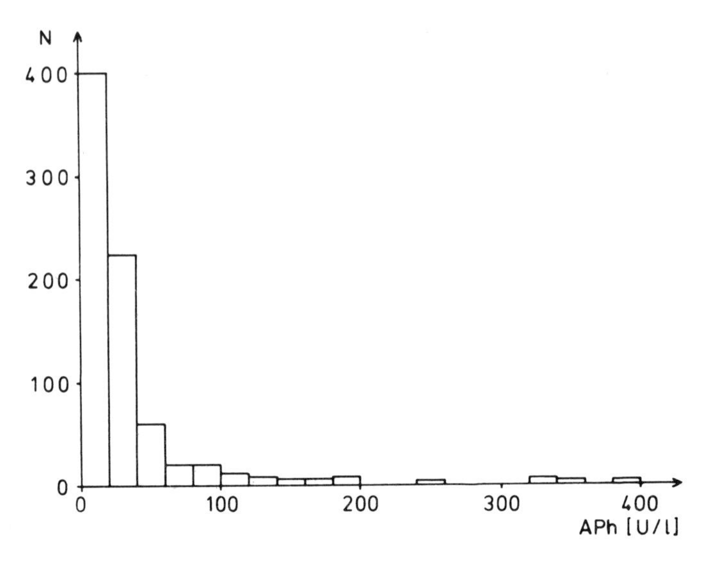

Figure 1. Histogram of alkaline phosphatase

Table 1 Extreme value control limits
Estimates of the limits into which 99 per cent of unselected laboratory results for some common tests are expected to fall.

substance	*0.99 limits*		*number*[a]	*units*
	lower	*upper*		
GOT	2	380	1000	U/l
LAP	5	70	785	U/l
APh	7	250	784	U/l
LDH	50	1200	783	U/l
Bil	0.2	15	992	mg/dl
Ca	3.5	6.7	719	mval/l
Gluc	40	530	560	mg/dl

[a] The number of assays on which each estimate is based

TREND CONTROL

If a test is made and the result is found to be in the pathological range and the test is repeated some time later a large difference between the two results could occur because the patient had returned to the normal state. In this sense the difference between correctly analysed results can be time dependent.

In our laboratory, abnormal test results are mostly repeated every day so that the time difference between successive tests is sufficiently small for the time dependence to be ignored when estimating the trend control parameters. The scatter diagram, line of regression and control limits of two successive results are presented in Figure 2. They are the first (X) and the second (Y) analysis of alkaline phosphatase, each pair being for the same patient. Table 2 contains examples for the limits of other serum and blood components.

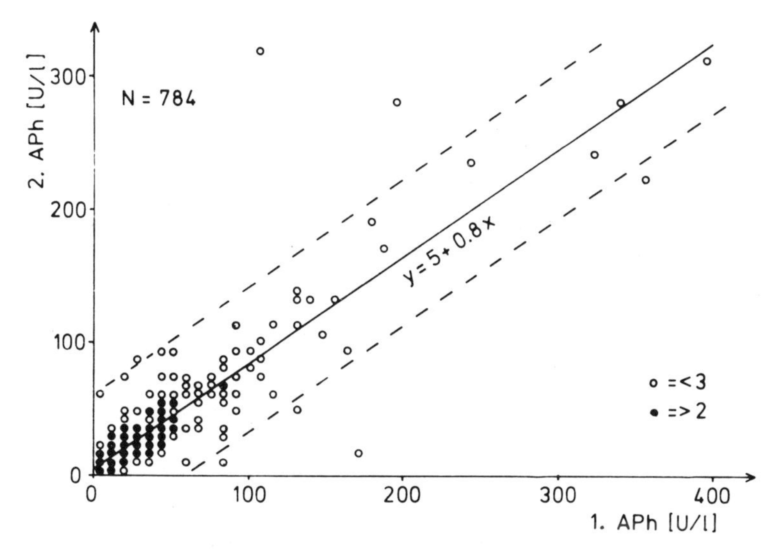

Figure 2. Scatter diagram, line of regression and control limits for two successive results of alkaline phosphatase

- ● based on mean of three or more assays
- ○ based on two or fewer

The limits are obtained in this way: First a linear regression (Y=A+BX) is computed. The two lines parallel to the line of regression are drawn, within which 0.99 of the points fall. If they are normally distributed results, one should compute the standard error of the estimate (0.997 of the points will fall within a region bounded by two lines drawn parallel to the line of regression at a vertical distance of 3 standard errors from it).

PATTERN CONTROL

There is a correlation between some laboratory results for a variety of reasons. If a correlation exists, one result can be controlled using the other result. A simple case would be the repeating of a test, possibly on a different analyser, in order to compare the two results. Most important are the relations between results, which change together in the course of disease: GOT and GPT during hepatitis, alkaline phosphatase and bilirubin during cholestatic jaundice, creatinine and urea by kidney insufficiency and so on. The limits for plausibility control are logical or statistical. The statistical limits enclose 0.99 of the value pairs. An example for pattern

Table 2 Trend control limits
The relation between two successive results for a given test on the same patient

substance	*correlation coefficient*	*number*[a]	*linear regression* $Y = A + B X$		*0.99 limits*		*units*
			B	*A*	*lower* *C*	*upper* *D*	
GOT	0.69	1000	0.51	9.6	60	210	U/l
LAP	0.87	785	0.76	3.4	23	21	U/l
APh	0.89	784	0.81	4.8	52	58	U/l
LDH	0.76	783	0.73	49	360	650	U/l
Bil	0.91	992	1.03	-0.1	4.6	6.1	mg/dl
Gluc	0.76	560	0.65	41	100	200	mg/dl
Ca	0.72	719	0.71	1.4	0.7	1.1	mmol/l

a The number of patients on which the parameters are based

control is shown in Figure 3. The limits are obtained in a similar way to those for pairs of results of the same test.

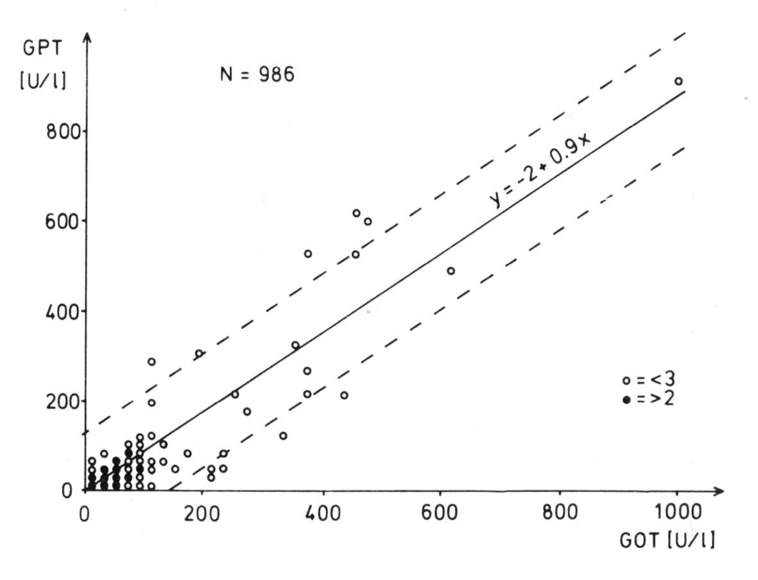

Figure 3. Scatter diagram, line of regression and control limits for GOT against GPT. Each pair of results is for the same patient

DATA COMPLETION

Plausibility control calls attention to laboratory results lying outside the ordinary range. The task of the laboratory doctor is to react to this information. Often the reaction is uniform: the laboratory doctor orders a repetition or a completion. A predictable reaction can be programmed. The computer orders repetition or completion using the typewriter in the respective laboratory. This way is short and very quick.

STEPS OF PLAUSIBILITY CONTROL

Plausibility control should be applied as early as possible in the processing of laboratory results. The earliest time is the moment when the results are received from the instruments. Which methods of plausibility control can be used depends on the instruments. On one-channel analysers only extreme value control and trend control is applied. The control limits for the extreme value control are mostly given by the measurement range. On multichannel analysers pattern control is also used. This plausibility control at the instrument is the first step.

The second step is plausibility control over all results of one day. This second step is realised before printing the summary report for the ward. All results are available. This is the best time for logical and statistical pattern control.

RESULTS

How does the laboratory profit from the plausibility control? Our laboratory produces 5000 to 8000 results per day. The plausibility program finds about 30 patients with remarkable laboratory values. 20 to 30 results are repeated. In this way one or two errors are found per day. This number seems little. But over a long time this method is effective and gives more security to the laboratory data.*

REFERENCES

1. Schmulling, RM, Graser, W and Eggstein, M (1975) In *Methoden der Informatik in der Medizin* (Ed) PL Reichertz and G Holthoff. Springer Verlag, Berlin, Heidelberg, New York. Page 159
2. Guder, WG (1977) *Med. Welt, 28,* 1249
3. Haeckel, R (1975) *Qualitatssicherung im medizinischen Labor,* Deutscher Arzte-Verlag, Koln. Page 117
4. Bessey, OA, Lowry, OH and Brock, MJ (1946) *J. Biol. Chem. 164,* 321
5. Karmen, A et al (1955) *J. clin. Invest. 34,* 126
6. Karmen, A (1955) *J. clin. Invest. 34,* 131
7. Wroblewski, F et al (1956) *Proc. Soc. exp. Biol. Med. 91,* 569
8. Nagel, W, Willig, F and Schmidt, FH (1964) *Clin. Wschr. 42,* 447
9. Wroblewski, F and La Due, JS (1955) *Proc. Soc. exp. Biol. Med. 90,* 210
10. Barnauch, D et al (1975) *Z. Klin. Chem. Klin. Biochem. 13,* 101
11. Jendrassik, L and Grof, P (1938) *Biochem. Z. 297,* 81
12. Ludwig, U and Eggstein, M (1973) In *Optimierung der Diagnostik* (Ed) H Lang, W Rick and L Roka. Springer Verlag, Berlin, Heidelberg, New York. Page 251

* Dr R Dybkaer of Copenhagen points out that delaying the reporting of some 20 'remarkable' but analytically correct results in order to detect one or two laboratory errors could be a dangerous procedure. Perhaps the solution is to report such tests immediately but emphasising to the clinicians that the results are provisional and the tests are to be repeated. Editor.

A MODEL TREND ANALYSIS IN CLINICAL CHEMISTRY BASED ON TWO VALUES

Dr A J Porth,
Medizinische Hochschule Labordatenverarbeitung (81.30),
Karl-Wiechert-Allee 9, D 3000 Hannover 61

INTRODUCTION

A statistical analysis of all results of biological parameters processed by the laboratory computer system of the Medical School, Hannover during one year led to a practicable means of plausibility control by means of trend analysis.

THE METHOD OF TREND ANALYSIS

Population: In a laboratory measured values for patients of a hospital

Given: V_0 previous value–value of a biological parameter (e.g. chloride in serum) (time = 0)

V_t later value–value of a biological parameter (time = t)

Definitions: 1. Difference trend

$$D_t (V_0, V_t) = V_t - V_0 \text{ (t = 1, 2, 3, . . . days)}$$

2. Percentage trend

$$P_t (V_0, V_t) = \frac{V_t - V_0}{V_0} \quad 100 \text{ (t = 1, 2, 9, . . . days)}$$

Constructing the trend-distribution:

Using V_0 and V_t belonging to the same patient, construct histograms of P_t and D_t, one histogram for each time range t (PA histogram–see Figure 1).

Construct similar histograms for V_0 and V_t not belonging to the same patient but combined at random to give a probability distribution for the subset of random sample identity errors (IE histogram–Figure 1).

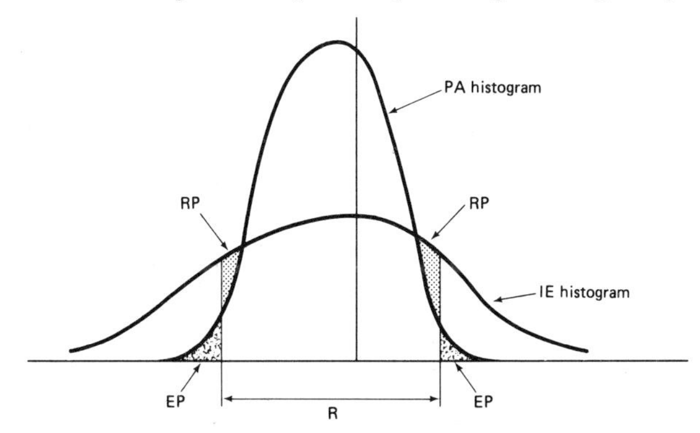

Select a decision range R to optimise the RP to EP ratio.

The probability space EP shows analytically correct results which would be queried on the basis of the deicsion interval R.

The probability space RP shows those results that are in error but would not be detected on the basis of using the selected R.

Main Results:

1. $EP = 5\%$ $\qquad RP \approx 10\%, \ldots, 60\%$

2. The expansion of this method to blocks of biological parameters (SMA 12, enzymes) leads to $RP \approx 70\text{-}80\%$ (with $EP = 5\%$)

DAILY PLAUSIBILITY CONTROL

The results of the investigation led to a daily check procedure supplying lists as shown in Table 1. The trend is presented in figures and in a graphical way to make it easier to pick up immediately the outstanding events.

Table 1. Examples of actual trend analysis

Testcode	*Previous results*	*Todays incoming results*	*D-Trend a*	*	*P-Trend a*
GXXXXXXXXX, Dietmar	15.0		2.63-2310 24B		
	010052	020053			
	1/07	2/07			
UNA	23	68	d		32 p+
UCL	12	80	d+		32 p++
	010053	020049			
	1/07	2/07			
XKCL	162	305	d++		55 p
	010076	010054			
	1/07	2/07			
SNA	140	121	d+		22 p--
SK	4.46	10.75	d+++++		23 p+++++
SBIL	64	21	d-		25 p-
SCA	2.59	2.11	d-		33 p-
SIP	1.07	1.97	d+		32 p+
SGLU	4.8	82.5	d+++++++++		22 p+++++++++
SKRE	36	133	d		35 p++++
SXXXXXXXXX, Petra		31.0	8.50-1520 15		
	290280	020172			
	29/08	2/08			
SLDH	191	391	d		24 p+
SCK	17	451	d+++		15 p+++++++++

a positive trend +
negative trend - (The number of signs is based on the values of tolerable trends evaluated by the author's study)
* First figure 10 = RP-value % of the D-Trend; Second figure 10 = RP-value % of the P-Trend } with EP ≈ 5%

CONCLUSION

The daily plausibility control procedure with the following check of errors helps to detect sample crossovers and other identity errors of patients with different result pattern. Investigations for reidentification procedures have now to be started.

THE COMPUTER IN A MICROBIOLOGY DEPARTMENT

I. Phillips MA, MD, BChir, MRCPath
Professor of Medical Microbiology
St Thomas's Hospital Medical School, London SE1 7EH

INTRODUCTION

The computer facilities in the Department of Microbiology, St Thomas's Hospital, are part of a much larger system based on a central Rank Xerox Sigma 6 computer serving many parts of the hospital, including patient registration and other pathology departments. It is the purpose of this paper to describe the uses to which the results obtained in the Department of Microbiology can be applied, other than the basic one of supplying a report on a specimen from an individual patient. These applications include aspects of laboratory and patient management, and have constituted some of the major benefits of the system.

The system used in the Microbiology Department is an extension of that developed at the Charing Cross Hospital and described by Andrews and Vickers (1). It will be described in detail elsewhere (2, 3) and it will suffice here to state that it uses an optical mark reader (OMR) system, and printed forms are used by technicians to book in specimens, (the computer allocating a laboratory number and printing out registration information on the patient) and as a work sheet. The forms are therefore large, with sections for recording all except the most unusual of the individual tests performed on any sample. The technician also records, in the final result section, the organisms isolated and their antibiotic susceptibility. The completed document is then fed into an OMR in the department and facsimile reports are scrutinised, edited and released on line on a visual display unit (VDU) in the department by laboratory doctors.

APPLICATIONS OF THE SYSTEM

The main objective of the system is to produce a report on an individual sample from an individual patient. Some aspects of this are worthy of discussion here, as they involve a use of individual laboratory results in a way not hitherto reported for a diagnostic laboratory.

Automatic Validation of Reports

The first advantage arises from the fact that the OMR document records almost all the work done by the technician so that both his test results and conclusions go into the computer. This means that conclusions on the identification of the organism can be checked by the computer against individual test results. We use only a few striking properties of individual organisms for this validation, but it is clearly possible to apply complete taxonomic systems should this be felt desirable. For example a report that *Proteus mirabilis* had been isolated would only be issued if the organism with a characteristic swarming colonial morphology had been shown to split area rapidly, and was resistant to colistin and tetracycline if antibiotic sensitivity tests had been performed. If results were atypical the report would only be released when the scrutinising pathologist had satisfied himself that they were correct. The major importance of this method is that many reports can be released automatically–for example samples for the Department of Genitourinary Medicine investigated for *Neisseria gonorrhoeae* are released whether positive or negative, the computer checking that if *N. gonorrhoeae* is to be reported, it is an oxidase positive, Gram-negative coccus that fluoresces with a specific serum, and which ferments glucose, but not maltose, lactose or sucrose if fermentation tests have been performed. The computer further checks that appropriate media have been inoculated and cultures examined after 24 and 48 hours incubation. All this means that the laboratory doctors can concentrate at scrutiny on detecting clinical problems on which they should act. This release of medical staff from the chores of routine scrutiny of reports has been most valuable.

LABORATORY MANAGEMENT

The production of statistical returns on work performed in the pathology department is time consuming, and we have used our computer facility to produce as many of them as possible. The simplest are the counts of requests received which form a major part of the returns required by the Department of Health and Social Security (DHSS). More complex and more rewarding has been the application of the Canadian Schedule of Unit Values (4). This has meant that we can now produce not only a count of requests, but also an accurate assessment of the total work done and can calculate real test request ratios, highly elusive under the system at present used for DHSS returns. This is of great value internally, as for the first time we can predict accurately the effects of extensions in clinical services, as well as being able more accurately to detect changes (usually increases) in workload month by month. Such information can be applied directly to laboratory budgeting.

The other aspect of laboratory management that has been helped by the advent of the computer can generally be referred to as quality control. Clearly we are now able to check that reports are soundly and correctly based on test results, and this helps to avoid the problem of the organism that constantly changes its identity in series of investigations, to the confusion of the clinician. However, with some specimens we can go further than this, and we have developed the system particularly with reference to cultures for the gonococcus from the Department of Genitourinary Medicine. In that department special OMR forms are used as request documents and a record is made of the results of Gram stains on the original sample of pus or exudate. The computer produces week by week a record of the correlation between these results and the results of culture. If culture results deviate substantially from our normal proportions for the specimens of the varying types, we investigate laboratory and clinic procedures and have been able to detect, for example, faults in our selective medium. This has proved a much more sensitive indicator of failures than have our standard microbiological quality control measures of inoculating the selective medium with laboratory-maintained strains of gonococcus. The method may well be applicable to other types of specimen.

A final aspect of laboratory management is control of the work flow. For the first time we can check regularly on the overall progress of samples through the laboratory and on samples on which investigations should be complete but have not yet been reported. We can also check readily on patterns of workflow—the timing of requests for blood cultures and the production by particular clinics of large volumes of work.

PATIENT MANAGEMENT

The computer is able to contribute to the management of the patient in several ways.

First, it can cumulate results for an individual patient from one or more departments to provide a constantly updated record of the patient's progress. The long-term aim for the St Thomas's system is that microbiology reports will be integrated into a general patient record, but as this has not yet been done, we can only guess that the final system will prove of great value to the clinician.

The second aspect of the computer's usefulness to the individual patient is in the management of as yet undiagnosed infection. Our system produces regular listings of organisms isolated from each type of specimen so that we are better able to predict associations between organisms and disease. It also produces listings of antibiotic sensitivities of the various groups of organisms. In the past we were able to do this laboriously only for *Staphylococcus aureus,* and this resulted in an antibiotic policy for the treatment of staphylococcal infection, which in turn resulted in an improvement in the treatment of individual patients. We came to appreciate that sensitivity to antibiotics is ever changing, and not always in the direction of increasing resistance. With the help of the computer we now produce regular listings of organisms and their antibiotic sensitivity so that we are able to establish and follow changes in sensitivity patterns of all antibiotics for a wide variety of organisms. We can now tell the clinician what are the chances, for example, of resistance to tetracycline among pneumococci or *Streptococcus pyogenes* or to ampicillin or sulphonamides in *Escherichia coli,* or to gentamicin among organisms from blood cultures, in current clinical material, so that he can choose an agent that is likely to succeed in patients who are acutely ill and as yet microbiologically undiagnosed.

This information is not only useful within St Thomas's Hospital, but with reservations to others, especially those who have no facilities for constant surveillance of this kind.

We are in the process of developing a more elaborate system to assist the early management of patients with septicaemia. This involves the computerisation of laboratory records on 1000 cases of septicaemia, which we have kept over the past eight years. The aim is to deliver to a clinician, suggestions on likely pathogens, their likely antibiotic sensitivity, and the likely response to appropriate therapy of patients in various clinical groups.

The other area in which the computer is invaluable is in the control of infection, particularly hospital-associated infection. In the past it was the function of the Infection Control Officer to collect information from the wards. The accumulation of statistics on only one type of infection, post-operative wound sepsis, proved so time-consuming that we were quite unable to study other kinds of infection, or indeed to do anything about the infection that we documented. This system has now been replaced by a computerised, laboratory-based system. We continue to exhort our clinicians to send samples to the laboratory from all cases of infection and over the years have found increasing, though never complete, compliance. The computer now produces lists of organisms with their antibiotic sensitivity patterns which are useful in preliminary assessments of possible outbreaks, and associates them with wards, departments, and operating theatres. As yet we have not been able to produce long-term statistical analysis, for example sepsis rates of each type of operation, or for an individual department, but the system is perfectly capable of this.

Assistance in infection control promises to be another of the major benefits of the system.

CONCLUSIONS

The development of the computer system for microbiology at St Thomas's Hospital, making as much use as possible of all the results obtained in the department has been produced only by sustained team effort. It has been particularly important to have systems analysts who were prepared to familiarise themselves with all aspects of our work and laboratory staff who were prepared to define their activities in minute detail.

Although the system is not yet fully developed, it has been in operation in microbiology for about three years. From the clinician's viewpoint, looking for results on an individual patient, the system has as yet produced only what he had before, with the development of integrated reporting still to come. In the laboratory, however, it has made a major contribution to the quality of our work, and to the orientation of medical staff towards clinical problems, providing analyses of results that are invaluable in the management of individual patients and in the control of hospital-associated infection.

ACKNOWLEDGEMENTS

The system described was developed by a team made up of laboratory doctors and technicians, notably Mr R Lynn and Miss E Rice from the Department of Microbiology, and members of the staff of the Department of Computing Science, especially Mr J M F Davidson and Mr K Williams. I am grateful to them for their enthusiasm and cooperation.

REFERENCES

1. Andrews, HJ and Vickers, M (1974) *J. clin. Path*, *27*, 185
2. Williams, KN and Davidson, JMF (in preparation)
3. Williams, KN, Davidson, JMF, Lynn, R, Rice, E, and Phillips, I (in preparation
4. Canadian Schedule of Unit Values of Clinical Laboratory Procedures (1976)

USING HAEMATOLOGY RESULTS:

A COMPUTER-BASED FAMILY PRACTITIONER ADVISORY SERVICE

Dr M.K. Alexander Consultant Haematologist
M.T. Corbett and R. Reed
Warwick Pathology Laboratory

INTRODUCTION

In considering the possible ways in which a computer might be profitably used in the clinical laboratory, relatively little attention has been given to its power to sort and order facts for direct use in community health care. The organisation of the health service in the United Kingdom, where in most areas the family practitioner has direct access to the services of the local clinical laboratory (including specialist advice), provides a particularly favourable environment for the employment of the computer to this end. The following communication is an account of the introduction of an advisory service for practitioners which is effectively dependent on the use of a laboratory minicomputer.

Warwick Pathology Laboratory, in addition to the service which it provides for hospitals in the district, deals with requests for work in all branches of clinical pathology from more than 100 family practitioners grouped into 40 practices. Specimens are either collected in the practices and brought to the laboratory by district courier transport service or taken by laboratory staff at outpatient clinics. The resulting workload represents about 25 per cent of the total laboratory workload for haematology of 200-250 specimens per day.

Test request information, identified by the patient's name and a two digit practice code is entered to a data processing system based on a Nova 1200 minicomputer of 32k core store with twin 2.5 megabyte discs. Results of blood counts performed on a Coulter Counter Model S are entered to the system on-line and other data, including differential leucocyte counts, by means of visual display units situated at the laboratory bench. Lists of abnormal results are produced at intervals for editing prior to the preparation of individual reports.

A system of selective cumulative recording is employed to store a proportion of reports. A full description of the principles involved has been given elsewhere (1), but briefly, the stimulus for opening a cumulative file is an abnormal result obtained from a patient hitherto not on file. Once opened, all further results from the same patient (identified by name and location) are added to the file which remains on disc until no fresh information has been added during the past 14 days, when it is automatically deleted unless it has been designated as permanent, in which case it is retained indefinitely at the discretion of the haematologist.

As with any filing system not based on an identifying characteristic unique to the individual, difficulties due to similarities of names occasionally arise. The size of the problem is, however, not as great as might be imagined: the largest practices concerned in the service number no more than about 8,000 patients and practitioners are moreover encouraged to use a six-character coded form of the patient's address in the request form field otherwise allocated to the hospital unit number.

On being made permanent, every file is given a diagnostic code and an 'activity' code. As regards the latter: '1' indicates that action on the patient is called for from the laboratory, '2' that action is awaited from the practitioner (e.g. in the form of further investigations as recommended by the haematologist) and '3' that no action is required. The activity code '2' is temporary and reverts automatically to '1' after 28 days.

The haematologist is presented every week with a list of files with activity code '1' for his attention and appropriate action. This may consist either of investigations which he initiates personally or of a reminder to a practitioner that further measures on his or her part had been recommended on a previous occasion.

Lists of the results (cumulative when present) for all patients on file, whether temporary or permanent, are printed for each surgery every fortnight. Where it is thought useful, the lists are annotated with details of other findings, indications for further tests and advice as to management. The availability of the list of activity code '1' files at this point greatly facilitates the task.

Used in this way the fortnightly lists serve as a valuable method of: firstly, reviewing patients from whom results have been obtained over a long period: secondly, of providing a consultative service for investigation and treatment: thirdly (and most importantly) as an *aide-memoire* both to haematologist and practitioner in cases where the significance of a set of results may have been in the first instance overlooked.

The service in its present form has been well-received by the practitioners concerned. From time to time the participants are consulted for suggestions as to the extension or addition of facilities. Some possible developments—for example the inclusion of biochemical results—are temporarily inhibited by limitations of disc capacity. A useful recent addition has been the introduction of regular bulletins containing items of general interest, e.g. innovations or alterations in laboratory tests or news of local epidemics as evidenced by laboratory isolations.

Any attempt to provide a comparable facility without the use of an automatic data processing system would be beyond the resources of almost all clinical laboratories in the country. With such a system it becomes not only practicable but a rewarding operation for the participants.

As is generally the case, it is impossible in practice to quantify the effect of the service on such matters as earlier diagnosis or reduction in hospital admissions but, prima facie, any support of this kind which can be given to the family practitioner may be expected to improve the quality of patient care.

REFERENCE

1. Alexander, MK, Corbett, MT and Reed, R (1977) *J. Clin. Path, 20,* 356

Part 7

RECENT ADVANCES IN NUMERIC TECHNIQUES

A COMPARISON OF DISCRIMINANT METHODS

P Sandel and **W Vogt**
Institut für Klinische Chemie am Klinikum Grosshadern der Universität München

INTRODUCTION

Multidimensional statistical analysis of laboratory results has come very much to the fore in diagnostic research in laboratories. Problems associated with the mostly empirical valuation of data and the availability of computers on which time consuming algorithms can be implemented have led to this development. Major attention is focused on mathematical procedures that make it possible to assign laboratory findings to a diagnosis or more generally to a group of laboratory findings of equal relative importance.

Two procedures suited to this purpose are discriminance analysis and cluster analysis. The difference between discriminance and cluster analysis is that the former requires a group assignment of each element used to calculate coefficients of the discriminant function. It is evident that the result of a discriminance analysis depends on how accurately the respective grouping is reflected by the information contained in the data.

In addition, normal distributions as well as identical covariance matrices are assumed within the different groups in the case of a linear discriminance analysis (1). A further characteristic in the procedure is the large dispersion of the coefficients of the discriminant function in the case of small sample sizes. Further important prerequisites such as data quality will not be discussed here.

The aim of the present study was firstly to show up the influence of sample size and of distribution forms on the outcome of the discriminance analysis and secondly to discuss the significance of the choice of group assignment. The study was performed on data from patients with a tentative diagnosis of thyrosis. For the first part the test data was supplemented by simulated data. In this connection simulation techniques for multidimensional data will be discussed briefly.

The present work is part of a larger study, consisting of several publications, which has been carried out together with Dr Jungst of the endocrinological outpatient department of the second medical clinic of the University of München which provided both the clinical data and the corresponding diagnoses.

DATA

In the case of 322 patients with a tentative diagnosis of thyrosis the triiodothyronine (T_3), thyroxine (T_4), thyroid stimulating hormone (TSH) and TSH after thyrotropin releasing hormone test ($TSH_{TRH\ test}$) were performed. The results together with an initial diagnosis made without knowledge of the laboratory findings were transferred to an electronic data carrier.

METHODOLOGY

Influence of sample size and distribution form on discriminance analysis

For the investigation of the influence of sample size and distribution form of raw data on classification in discriminance analysis, a result of the second part of this study was used; namely, the knowledge of a group assignment satisfying the information content of the data.

To start off with, a linear and a quadratic discriminance analysis was carried out on actual test data using this group assignment. The advantage of the quadratic discriminance analysis is that neither normally distributed values nor identical covariance matrices need to be assumed (2). After each analysis the data were reclassified on the basis of the discriminant function obtained. In addition a cross validation was performed by means of a recursive 'hold-one-out' technique. Then simulated data with equal sample size and approximately equal distribution forms were generated and also used in a linear and a quadratic discriminance analysis. Further simulation runs were made using two thirds and one third of the initial sample size and finally on samples with less skew in the distribution.

The following method was used to generate simulated data. The first step was to determine the skewness of the distributions of the variables by using a maximum likelihood method recommended in 1951 by Cohen (3) in an iterative process to determine an α such that the value $z = \log(x - \alpha)$ and $z = \log(\alpha - x)$ respectively is practically normally distributed. Positive and negative excess, however, cannot be taken into account in this procedure.

For each group the results are determined separately. In the one-dimensional case it is sufficient to determine the mean value $\bar{z}$ and the standard deviation s_z of the data transformed with α. Then simulated data are to be obtained from normally distributed random variables r with mean $\bar{z}$ and standard deviation s_z by means of the inverse transformation $x = \exp(r) + \alpha$.

In the multidimensional case account has to be taken not only of the mean and variance of each variable but also of the dependence between the variables. For this purpose the covariance matrix C_z of the transformed values z was determined in addition to the vector of the alpha-values $A = (\alpha_1, \ldots, \alpha_n)$ and the vector of the mean values $\bar{Z} = (\bar{z}_1, \ldots, \bar{z}_n)$. The principal components (4) of the variances of the covariance matrix were now determined. This is equal to a determination of the eigenvalues l_i of the covariance matrix.

The simulation was then carried out in four stages:

1. Simulation of vectors R_o with $N(o, l_i)$-distributed independent components.
2. Generation of the dependencies by mapping the vectors R_o into R_d using the eigenvectors associated with the eigenvalues l_i.
3. Adjustment of the vectors R_d by the mean $\bar{Z}$ to form the vector $R' = R_d + \bar{Z}$.
4. Mapping of the vector R' into the original space using the $x_i = \exp(r_i') + \alpha_i$ for each component i.

Significance of prior knowledge of group assignment in discriminance analysis

To elucidate the significance of the group assignment in discriminance analysis, the values were grouped according to the clinical diagnosis made without inspection of the laboratory findings and a discriminance analysis was performed. Then a cluster analysis was carried out using a weighted difference measure recommended in 1963 by Ward (5). After evaluation of the dendrogram following the cluster analysis, a group assignment according to the clustering and independent of the diagnoses was made and a discriminance analysis was carried out.

RESULTS

The effect of sample size on the result of discriminance analysis was investigated with actual and simulated data (Figure 1). It can be shown that with either data the portion of correct reclassification as well as correct cross validation increases with decreasing sample size. This is the case for both the linear and the quadratic discriminance analysis. It is obvious that for a small sample size a deceptively good result is produced. This is probably due to the fact that for small sample sizes the total variability of the data is not taken into account. This is also the reason that generally the difference between correct reclassification and correct cross validation increases with decreasing sample size.

		real data			simulated data		
		N	2/3N	1/3N	N	2/3N	1/3N
linear discriminant analysis	% correct reclassification	84·6	88·4	90·7	77·0	81·4	87·4
	% correct crossvalidation	83·7	86·5	86·9	76·3	80·0	82·4
quadratic discriminant analysis	% correct reclassification	90·9	92·5	95·3	74·8	86·5	88·0
	% correct crossvalidation	88·4	89·7	86·9	70·7	81·9	84·3

Figure 1. The influence of sample sizes on the results of discriminance analysis. The rate of correct reclassification and of correct cross validation as well as their difference increases with decreasing sample sizes. The result of quadratic discriminance analysis is generally better than of the linear one.

The effect of distribution forms on the results of discriminance analysis was investigated solely with simulated data (Figure 2). For the simulation of the original data those α-values were used for which the values $z = \log(x - \alpha)$ and $z = \log(\alpha - x)$ respectively were largely normally distributed.

		A	B	C
linear discriminant analysis	% correct reclassification	77·0	86·0	90·0
	% correct crossvalidation	76·3	84·1	89·4
quadratic discriminant analysis	% correct reclassification	74·8	86·0	90·3
	% correct crossvalidation	70·7	84·7	88·2

Figure 2. The influence of simulated distribution forms on the result of discriminance analysis. The correct classification rates increase with better approximation to normally distributed forms while the difference between the results of linear and of quadratic discriminance analysis decrease. Distribution form: A = real data, B = approximately Gaussian, C = nearly Gaussian

For a first approximation of normally distributed forms the difference between the α-value and the data was increased fivefold while for a good approximation the difference was increased twentyfold.

This change towards normally distributed forms has a positive effect on the linear and the quadratic discriminance analysis. The correct reclassification as well as the cross-validation is markedly improved even after the first approximation.

Between linear and quadratic discriminance analysis the differences are notable. The rate of correct reclassification as well as crossvalidation is generally better with quadratic discriminance analysis. These differences are clear with real data and vanish if data changes towards normally distributed forms.

To investigate the effect of grouping, the data were sorted into nine groups according to the clinical diagnoses made without knowledge of the laboratory findings (Figure 3). The smallest group of 5 represents

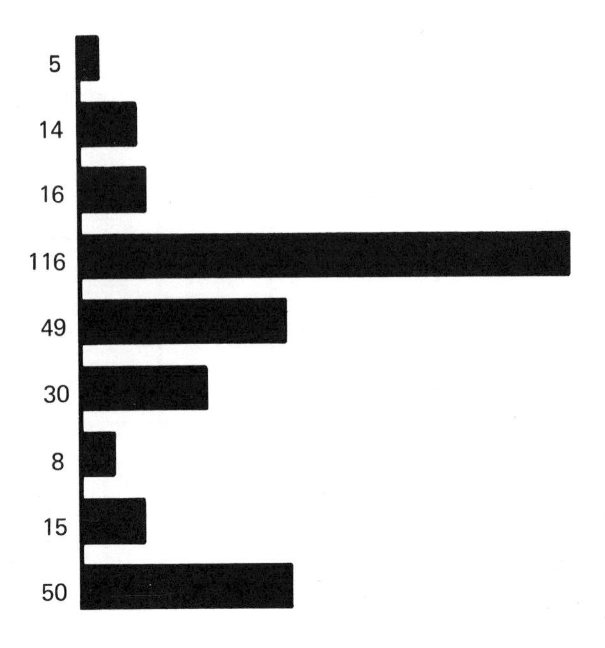

Figure 3. Grouping according to initial clinical diagnoses made without knowledge of the laboratory findings

the group of primary hypothyroidism while the largest group of 114 is that of bland diffuse strumae.

The result of the discriminance analysis is unsatisfactory for this group assignment (Figure 4). The correct reclassification rate is 41.2 per cent and 40.2 per cent for the linear and quadratic discriminance analysis respectively. Corresponding rates for the cross validation are 39.5 per cent and 37.5 per cent. In this context it is notable that the correct reclassification and cross validation is better for groups of smaller size by the quadratic discriminance analysis than by the linear one. On account of the large number of incorrect allocations it is easy to conclude that the group assignment did not reflect the information content of the data.

Therefore a cluster analysis was made. The final steps of this clustering are depicted as a tree in Figure 5. On the basis of this clustering we chose a division into five groups. A comparison of this new group assignment with grouping according to the diagnosis is represented in a diagnosis-cluster matrix (Figure 6) and shows that the two group allocations have virtually nothing in common: the sample sizes in the diagnosis-cluster matrix seem to be arbitrarily distributed.

After a comparison of the separate distributions of the variables obtained by the regrouping into five new groups with the distributions of the variables obtained from all sample sizes (Figure 7a, b) two points become immediately evident. Firstly the distributions show well defined forms and secondly there exist definite differences between the

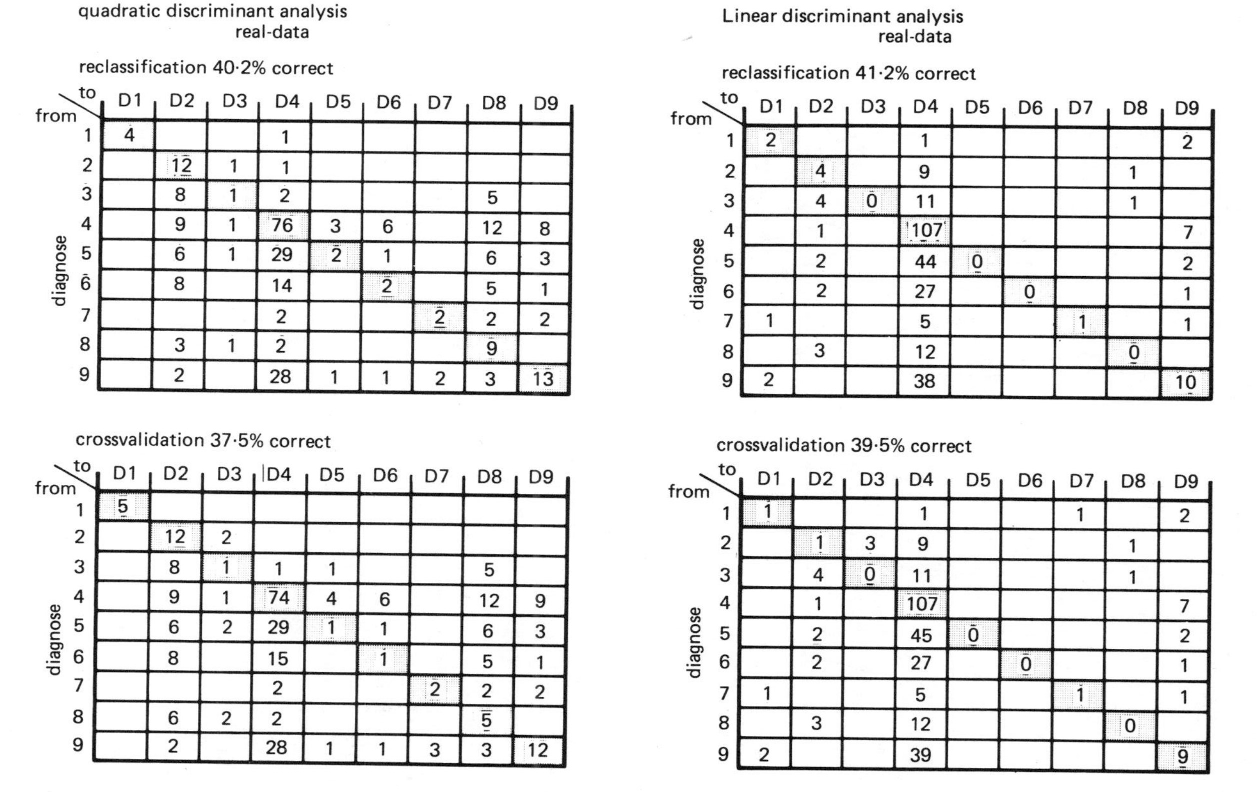

quadratic discriminant analysis
real-data

reclassification 40·2% correct

diagnose from \ to	D1	D2	D3	D4	D5	D6	D7	D8	D9
1	4			1					
2		12	1	1					
3		8	1	2				5	
4		9	1	76	3	6		12	8
5		6	1	29	2	1		6	3
6		8		14		2		5	1
7				2			2	2	2
8		3	1	2				9	
9		2		28	1	1	2	3	13

crossvalidation 37·5% correct

diagnose from \ to	D1	D2	D3	D4	D5	D6	D7	D8	D9
1	5								
2		12	2						
3		8	1	1	1			5	
4		9	1	74	4	6		12	9
5		6	2	29	1	1		6	3
6		8		15		1		5	1
7				2			2	2	2
8		6	2	2				5	
9		2		28	1	1	3	3	12

Linear discriminant analysis
real-data

reclassification 41·2% correct

diagnose from \ to	D1	D2	D3	D4	D5	D6	D7	D8	D9
1	2			1					2
2		4		9				1	
3		4	0	11				1	
4		1		107					7
5		2		44	0				2
6		2		27		0			1
7	1			5			1		1
8		3		12				0	
9	2			38					10

crossvalidation 39·5% correct

diagnose from \ to	D1	D2	D3	D4	D5	D6	D7	D8	D9
1	1			1			1		2
2		1	3	9				1	
3		4	0	11				1	
4		1		107					7
5		2		45	0				2
6		2		27		0			1
7	1			5			1		1
8		3		12				0	
9	2			39					9

Figure 4. Rates of reclassification and of cross validation by linear and quadratic discriminance analysis. The correct classification of elements of small groups is better by quadratic than by linear discriminance analysis.

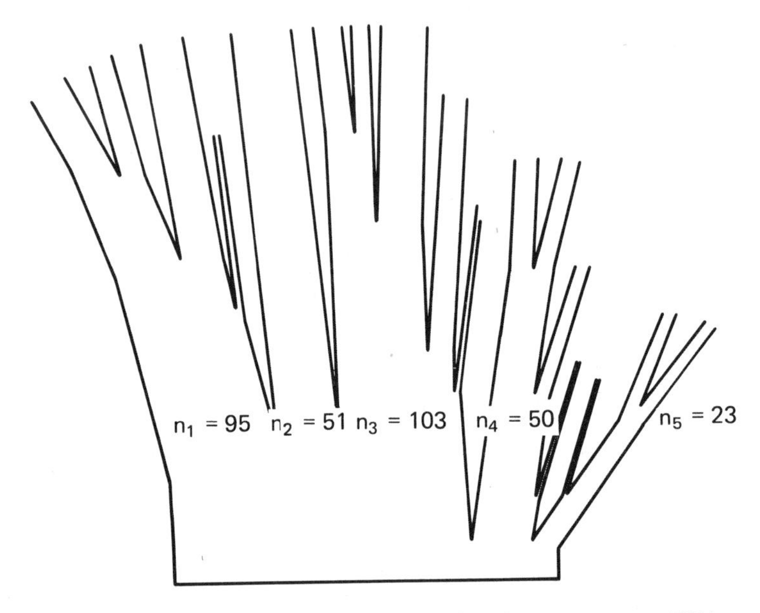

Figure 5. Tree formation after clustering using four parameters TSH_{TRH}-test T_3, T_4, TSH and TSH_{TRH}-test. Five groups were chosen according to the main trunks of the tree

distribution patterns of the individual groups. This is by no means an obvious result of a cluster analysis since for example a densely packed multidimensional meander would also be shown up as a cluster.

The pathophysiological meaning of the groups found will be discussed elsewhere (6). Only in conjunction with the inherent meaning of this grouping is an application of a cluster analysis and a subsequent discriminance analysis meaningful.

The result of the discriminance analysis after clustering is in accordance with the clear separation of the distributions, satisfactory: in the case of the linear discriminance analysis the portion of correctly reclassified cases is 84.6 per cent while 83.7 per cent are grouped correctly by cross validation (Figure 8). The corresponding values for the quadratic discriminance analysis are 90.9 per cent and 88.4 per cent respectively.

CONCLUSIONS

In discriminance analysis a too favourable result is obtained for small sample sizes. If the difference between reclassification and cross validation rates becomes too large the sample size must be increased. This may not be done, however, at the expense of the homogeneity of the data.

The effect of distribution forms should be taken into account in discriminance analysis. If data are skewly distributed or if covariance

	D1	D2	D3	D4	D5	D6	D7	D8	D9
Cluster 1	1		1	44	16	5	1		21
Cluster 2		2	3	20	8	5		4	6
Cluster 3		3	5	35	18	14	4	7	10
Cluster 4		9	7	13	6	6		4	3
Cluster 5	4			4	1		3		10

Figure 6. Diagnosis-cluster matrix. The diagnoses have nothing in common with clustering. They seem arbitrarily distributed

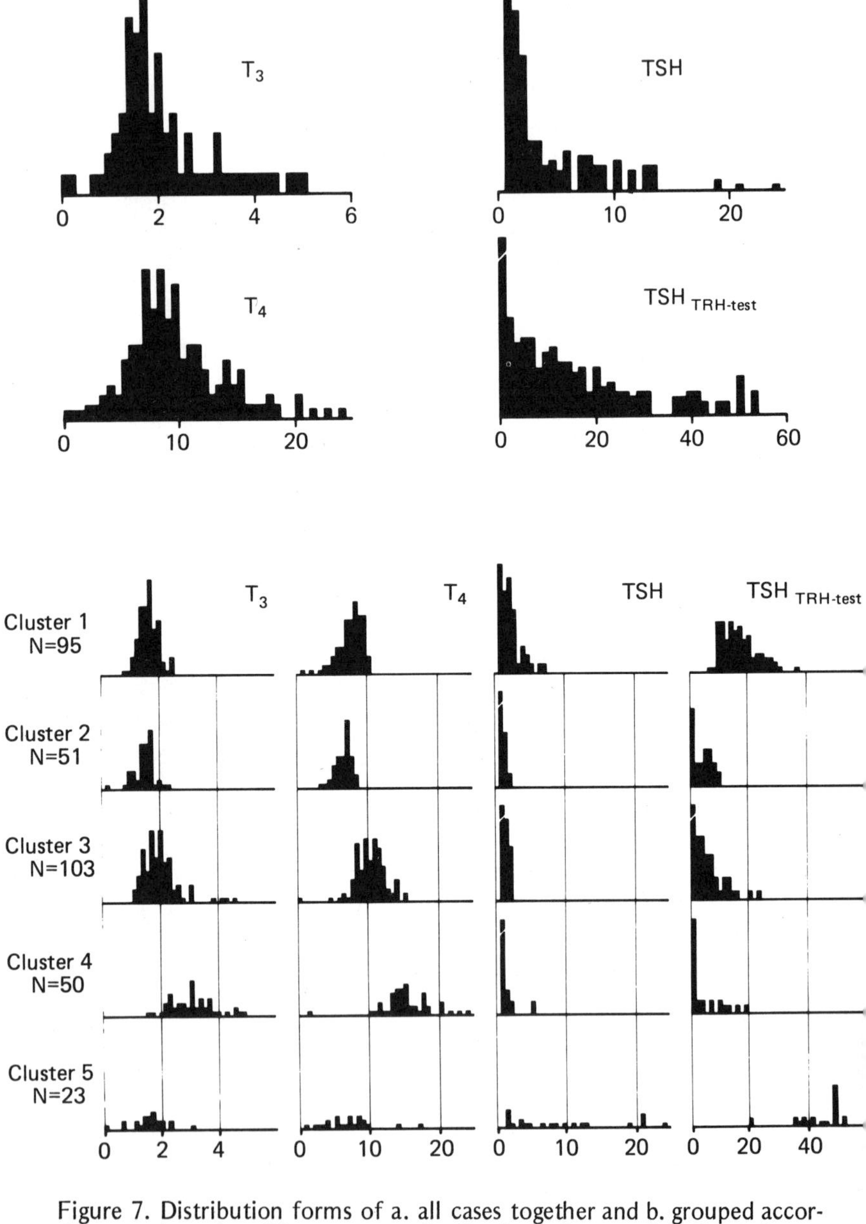

Figure 7. Distribution forms of a. all cases together and b. grouped according to clustering. The separate distributions show well defined forms. There exist definite differences between the distribution patterns of the individual groups.

Linear discriminant analysis

reclassification 84·6% correct

from \ to	C1	C2	C3	C4	C5
Cluster 1	79	1	13		1
Cluster 2		44	6		
Cluster 3	7	5	87	3	
Cluster 4	1		10	39	
Cluster 5	1			1	21

crossvalidation 83·7% correct

from \ to	C1	C2	C3	C4	C5
Cluster 1	78	1	14		1
Cluster 2		44	6		
Cluster 3	7	5	86	4	
Cluster 4	2		10	38	
Cluster 5	1			1	21

Quadratic discriminant analysis

reclassification 90·9% correct

from \ to	C1	C2	C3	C4	C5
Cluster 1	87	2	4		1
Cluster 2	2	48			
Cluster 3	4	4	90	4	
Cluster 4	1		7	42	
Cluster 5	1				22

crossvalidation 88·4% correct

from \ to	C1	C2	C3	C4	C5
Cluster 1	85	2	4		3
Cluster 2	2	46	1	1	
Cluster 3	5	4	89	4	
Cluster 4	1		9	40	
Cluster 5	1				22

Figure 8. Correct reclassification and cross validation rates by linear and quadratic discriminance analysis after clustering. The result is satisfactory. It is better by quadratic than by linear discriminance analysis.

matrices are not identical better results can be expected by the quadratic discriminance analysis than by the linear one.

The most important aspect is the correct grouping of the data. A cluster analysis on the laboratory data can be very beneficial. The grouping after an initial clinical diagnosis does not always reflect the information contained in the data.

REFERENCES

1. Kendall, M (1975) *Multivariate Analysis.* Charles Griffin and Co. Ltd., London, High Wycombe. Page 147
2. Welch, BL (1939) *Biometrika, 31,* 218
3. Cohen, AC (1951) *J. Am. Stat. Ass., 46,* 206
4. Anderson, TW (1958) *An Introduction to Multivariate Statistical Analysis.* John Wiley and Sons, New York
5. Ward, JH (1963) *J. Am. Stat. Ass., 58,* 236
6. Sandel, P, Vogt, W, Broda, S, Jungst, D and Knedel, M. accepted for presentation at the *Tenth International Congress of Clinical Chemistry* February 1978, Mexico

DEVELOPMENTS IN CURVE-FITTING PROCEDURES FOR RADIOIMMUNOASSAY DATA USING A MULTIPLE BINDING-SITE MODEL

P.G. Malan BSc, PhD, **M.G. Cox*** BSc, PhD,
Enid M.R. Long* MA, BSc, and **R.P. Ekins** MA, PhD
Sub-Department of Molecular Biophysics,
Department of Nuclear Medicine, The Middlesex
Hospital Medical School, London W1N 8AA, and
**Division of Numerical Analysis and Computing,*
National Physical Laboratory, Teddington, Middlesex TW11 0LW

INTRODUCTION

Several computer programs have been published for processing radioimmunoassay (RIA) data, and they may be divided into two broad categories on the basis of the methods used to fit curves to the standard or calibration points. Curve-fitting procedures will be referred to as 'model-based' when they rely on a mathematical model representing the fundamental physico-chemical kinetics of the reagents employed in the assay system. In contrast, those methods which disregard the physico-chemical basis of RIA, and which employ empirical models to fit the experimental response points yielded by assay standards, will be referred to as 'data-based' or empirical fitting procedures. Any 'lack of fit' by the latter procedures is usually attributed to a deficiency of the empirical curve-fitting procedure used. On the other hand, those methods using theoretical models based on the Law of Mass Action have been employed moderately successfully, but they suffer from the simplifying assumptions which have been necessary in order to implement these simple model-based curve-fitting procedures on relatively small computers or calculators.

We shall compare some of the more common model-based and empirical curve-fitting procedures which have been used, and we shall also describe the results obtained with a multiple binding-site curve-fitting model. The latter sophisticated model will successfully fit a wide range of assay data, and it also provides estimates of binding-site concentrations and the values of the respective equilibrium constants present; we have used the latter reagent constants to refine further the assay conditions by employing computer optimisation techniques (1).

MODEL-BASED CURVE-FITTING PROCEDURES

One Binding-site Mass Action Model

The solution to the Law of Mass Action for a single ligand and a single binding site has the form of an hyperbola (2), when either the bound-to-free (B/F) or the free-to-bound (F/B) ratio is plotted against ligand concentration. The latter curve (shown in Figure 1a of reference 2) yields an approximately straight line dose-response relationship, and if sufficient labelled ligand is added to 'saturate' the binding site present, the curvature seen at the low end of the standard curve may be largely or completely obscured. The point at which the standard curve intercepts the ordinate is usually referred to as $(F/B)o$.

The slope of the asymptote to the F/B *vs* ligand hyperbola has a slope equal to the reciprocal of the binding-site concentration (i.e. $1/q$). If only one binding-site is present, and the labelled ligand saturates this site, then the experimental curve and its asymptote lie very close together and would be indistinguishable experimentally. Hence, under these conditions, the equation of the standard curve may be approximated by a straight line:

$$F/B = p/q + (F/B)o \tag{1}$$

where p is the standard ligand concentration, and q is the binding-site concentration in the assay system. Similarly, it is equally legitimate to plot total-to-bound (T/B) ratio against ligand, since:

$$T/B = (F + B)/B = F/B + 1 \tag{2}$$

which yields the same form of dose-response that is displaced only by unity from the F/B curve.

Labelled ligand which is 'non-specifically' bound may be modelled as a second species of binding-site of low affinity and high capacity in this assay system. Non-specific binding (N) has the effect of causing curvature of the F/B (or T/B) response curve at high ligand concentrations (2). A simple conventional method of correcting for this effect is to subtract the non-specific term (N), which frequently re-linearises the dose-response curve:

$$(F + N)/(B - N) = m.p + (Fo + N)/(Bo - N), \tag{3}$$

where m represents the slope of the straight line.

The Logit-log Model

The two-parameter logit-log model advocated originally by Rodbard and Cooper (3) has the form:

$$\text{logit}\,(b/bo) = \log\left(\frac{b/bo}{1 - b/bo}\right) = -\text{r}\,.\log p + \log a,$$

where $-r$ is the slope of the line, $\log a$ is the ordinate intercept, and b is the fraction bound minus the non-specific term ((B – N)/T). Equation 4 may be rearranged:

$$\log\left(\frac{b}{bo - b}\right) = \log\,(ap - r),$$

and after taking antilogs and regrouping terms:

$$1/b = pr/(a.bo) + 1/bo,$$

or, (5)

$$T/(B - N) = \text{m}.pr + \{T/(B - N)\}\text{o}.$$

Following from the relationship in equation 2, the simple Mass Action model (equation 3) and the log-logistic model (equation 5) are thus formally identical *when the slope of the logit plot is unity,* i.e. when $r = 1$. However, the logit-log model may *empirically* accommodate data which do not provide a straight line dose-response relationship on a F/B *vs* ligand plot for values of r which are greater than or less than unity, as is shown in Figure 1. In this way, the logit-log model is capable of providing

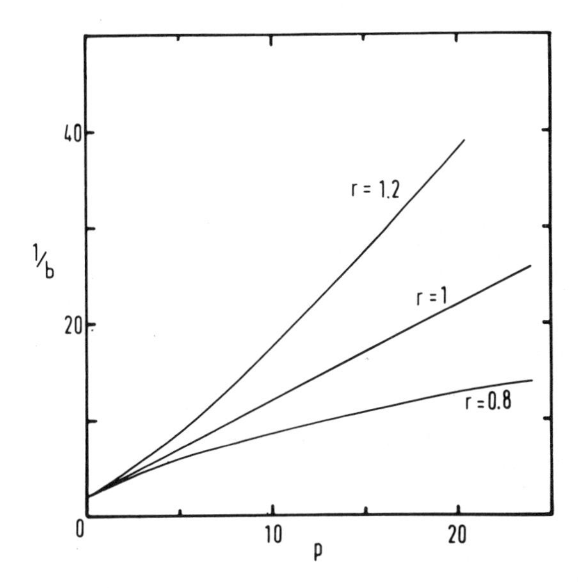

Figure 1 The reciprocal of fraction bound is plotted against ligand concentration. The curves were produced by substituting values of r = 0.8, 1.0 and 1.2 in the power-function (equation 5), which was derived from the logit-log expression. The constant m was set = 1.

a reasonable fit to the data, even when poor, or incorrect estimates of (F/B)o and non-specific binding are used.

Several groups (4-9) have published other methods of curve fitting which rely essentially on one or other of these two models. However, if an assay system does not conform to the conditions where a single species of binding site is saturated with labelled ligand, and the reactions are at or near equilibrium, then the above models may fail or require some modification.

EMPIRICAL CURVE-FITTING PROCEDURES

Extensions to the Simple Model

The power-function (equation 5) derived from the log-logistic model is not obtained from the kinetic equations describing the equilibria, so modifications of the latter model are considered to be further extensions of the empirical approach. An improvement to the simple model was first introduced by Healy (1) and was later adopted by Rodbard and

Hutt (11): the experimental estimates of Bo and non-specific (N) are subject to assay imprecision, so the fitting procedure is used to estimate Bo and N, as well as the slope and intercept, as fitted parameters. This complicates the computational methods, but a general fit to a class of S-shaped curves is thus obtained. In a similar way, we modified the Taljedal and Wold(8) algorithm to accommodate a wider range of assay data (12).

Spline Functions

Cubic spline functions are more flexible and are thus more generally applicable to RIA curve-fitting than are polynomials. A cubic spline is a function which consists of a number of cubic polynomial arcs or segments bounded by points known as knots; at the latter points, the neighbouring arcs smoothly join each other and have the same value, slope and curvature. The knot positions are selected initially according to an appropriate strategy, and the spline is then fitted to the standard points using a linear algorithm. Our approach has been different from that of Marschner et al (13) who have employed the technique (14) where a knot was placed at every experimental point, and suitable weighting of the points, plus a superimposed smoothing function, was used in combination to yield a fit to the standard curve.

For the past four years, we have been using a numerically stable algorithm for least-squares cubic spline fitting, which was developed at the National Physical Laboratories (15). One or two knots are placed at appropriate points within the range of the data before the spline is used to fit the curve: the number of knots is always chosen to be, at most, equal to the number of fitted points minus five. Such a choice ensures that the number of degrees of freedom of the fit is less than that of the data, and a least-squares fit to the data is thus obtained, which may be either weighted or unweighted. Cubic functions tend towards plus or minus infinity in the ordinate (y) direction, and it is often difficult to obtain a satisfactory fit which is free from oscillations near the ends of the range of assay data plotted in the normal manner, where the data points are horizontal in the abscissa direction, e.g. in plots of per cent bound *vs* log (ligand), or per cent bound *vs* ligand. Accordingly, we have chosen to fit the curve as log (ligand) on the y axis, against per cent bound. The logarithmic axis spaces the data approximately evenly when the standards are prepared by doubling-dilutions.

The spline-fitting technique is carried out interactively, by allowing the operator to change the knot positions. The quality of the fit is assessed according to several criteria including minimising the residual sum of squares, examining the magnitude and number of sign-changes of the residuals, and the local and global behaviour of the fit itself. We have purposely chosen not to automate the fitting procedure, as we feel that the present method gives the operator a clearer understanding of the fitting procedure, and it also ensures that the operator assesses the criteria of 'goodness of fit'. In only a very few cases does the spline function yield an inferior curve-fit (16). These are the cases where a simple, straight line model would suffice, or where distortion of the curve occurs because of outlier data-points.

A MULTIPLE BINDING-SITE MODEL

In order to improve the generality of curve-fitting procedures, and because of the difficulty in assessing whether data points should actually be treated as outliers, we have examined the following multiple binding-site model which was originally formulated by Ekins et al (2):

$$1/R = \sum_{j=1}^{n} \frac{q_j}{p/(1 + 1/R) + 1/K_j} \qquad (+ K_{n+1} \cdot q_{n+1}) \qquad (6)$$

In this equation, R is free-to-bound ratio (F/B), j represents the species of binding site from 1 to n, while qj and Kj are the concentration and equilibrium constant of the jth site, respectively; p is the total ligand concentration present. Non-specific binding may be modelled as either the nth term in this equation, with a low affinity and high capacity, or as the combined $(n + 1)$th term as shown in parentheses in equation 6.

This formulation constitutes a non-linear implicit model (i.e. the parameters appear non-linearly): the response of R/B ratio cannot be expressed explicitly in terms of p, or vice versa, except in very simple cases, e.g. where $n = 1$. For curve-fitting purposes, the above model has been reexpressed in terms of transformed variables, and details of the technique will be published elsewhere. Satisfactory fits to a variety of RIA data have been obtained using a computer program written at the National Physical Laboratory in ANSI FORTRAN (17), which minimises a non-linear function subject to prescribed upper and lower bounds on the parameters, i.e. K's and q's. The program runs on our Mod. Comp. mini-computer, and only relatively few iterations ($\leqslant 200$) are normally required to achieve satisfactory convergence, even from poor starting values of the parameters. Table I shows the agreement achieved between the fitted parameter values and those obtained from the assays shown in the table, and similar results have also been obtained for aldosterone, arginine vasopressin, vitamin B_{12} and vitamin D assays. In all cases, the number of parameters which yielded the most reasonable fit has been selected, and the agreement is very acceptable when the limitations of the graphical Scatchard procedure are considered.

No objective criteria have been defined for assessing the superiority of one curve-fitting method over another. We have taken a representative standard curve from a routine assay, and ten experienced assayists were asked to draw the curve manually themselves, and then interpolate 19 points from the resulting F/B *vs* hormone plots to yield the hormone concentration. The data were pooled to obtain the mean and standard deviation at each point of interpolation. A generalised non-linear least-squares program (18) was used to generate curve-fits to the unweighted data using the curve-fitting algorithms described in the legend to Figure 2, and computer interpolation was performed at the same 19 points as were read off the standard curve manually.

The difference between each of the manual and computer interpolated ligand concentrations is plotted in Figure 2, against the standard ligand concentration. The difference for each curve fit is compared with the hatched area which shows the ±1 s.d. limits obtained for the manual fits.

Table I

Selected results from curve-fits to RIA data using the multiple binding-site program (17). Initial values for parameters were obtained by graphical Scatchard analysis, and only those combinations of parameters which yielded satisfactory fits are shown below. (hGH = human growth hormone; Ins = insulin; T4 = thyroxine; T3 = triiodothyronine; *r*T3 = 'reverse' T3; E3 = oestriol).

Assay		*hGH*	*E3*	*Ins*	*T4*	*T3*	*rT3*
No. of data points		10	8	11	6	6	8
No. of parameters		4	3	4	3	3	5
q_1	Initial	0.3	98	1.49	4.0	1.6	9.41
	Final	0.293	112	1.29	4.13	2.41	6.36
K_1	I	5.3	0.098	5.83	0.25	0.625	0.595
	F	5.40	0.0411	11.9	0.527	0.415	0.0771
q_2	I	0.8		25.4			58.8
	F	0.836		638			14.4
K_2	I	0.109		0.0061			0.0017
	F	0.103		0.0002			0.00696
$K_{n+1} \cdot q_{n+1}$	I		0.10		0.079	0.131	0.095
	F		0.0158		0.0827	0.115	0.0872
Residual sum	I	0.0456	16.8	4.52	–	–	–
of squares	F	0.0276	0.005	0.0148	0.0011	0.0308	0.00350
No. of function evaluations		6	14	24	9	8	44

The three-parameter multiple binding-site model, the quadratic fit and the spline fit is each considered to be reasonable, while the two 4-parameter, non-linear, empirical models are probably satisfactory. The unweighted logit-log and linear fits were unsatisfactory, even though the former results lay within the confidence limits for the ligand which were derived from the error relationship for this assay (i.e. $\pm \Delta H$ in Figure 2).

SUMMARY

A minicomputer has been used to develop and run programs for processing data from routine radioimmunoassays in this laboratory over the past four years. Points on the assay standard curves have been successfully fitted by cubic splines or, if appropriate, transformations are made to the data by straight lines. Not all assays can be accommodated by simple empirical transforms of the data intended to yield a straight line fit, however. Also, the flexibility of the spline function paradoxically constitutes a disadvantage, when uncertainty exists concerning the validity of particular experimental points which comprise the standard curve. Since the fundamental physicochemical reactions underlying RIA are well defined, a further approach is to fit the standard curves using the kinetic equations. The advantages of employing such a model-based curve-fitting procedure lie in its general applicability to a variety of assays, whilst the constraints imposed by the model enable outlier points to be identified more readily. In addition, the fitted parameters may be used

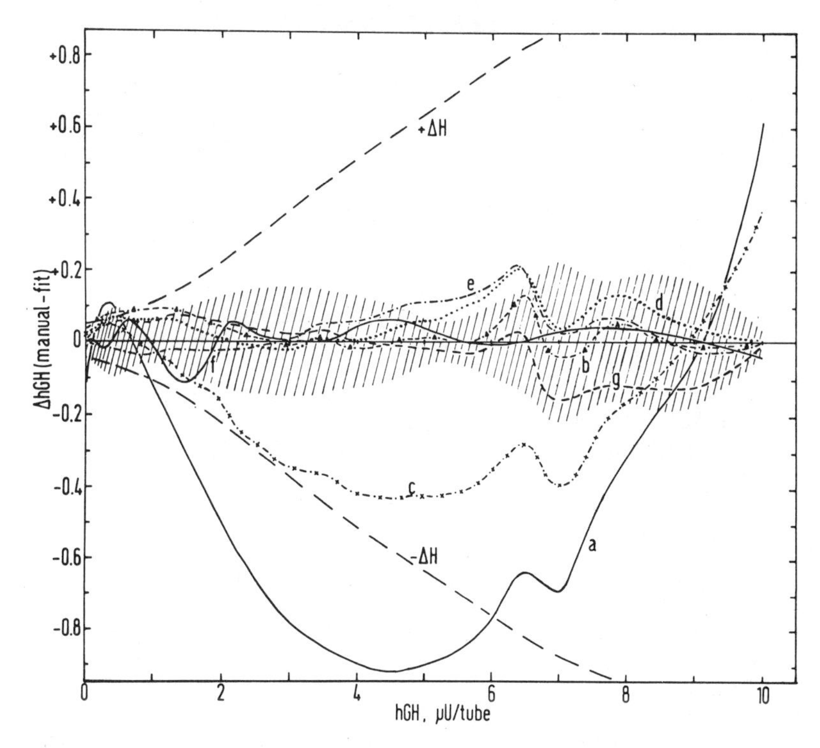

Figure 2 Different curve-fitting algorithms are compared with the mean fit to the data obtained manually by 10 individuals, for a human growth hormone assay. The differences between the fitted and manual results are plotted against standard ligand concentration, and the shaded region indicates ±1 s.d. of the manually fitted curves. The lines represent the following different curve-fitting algorithms: (a) a straight line, and (b) a quadratic (both with y = T/(B-N), and $x = p$); (c) the logit-log (4); (d) our non-linear hyperbolic function (12) (y = %B, $x = p$); (e) the Rodbard and Hutt 4-parameter model (10) (y = %B, $x = p$). The same data were also obtained for (f) the spline-fit (16), after adjustment of the knots, and (g) the multiple binding-site model, after alteration of the number of parameters, to obtain the most satisfactory fit.

for further refining the assay conditions by the use of computer optimisation techniques.

A range of assays has been examined in an attempt to assess the ability of a multiple binding-site model to provide satisfactory curve fits to different RIA data. The comparison procedures used, and a selection of the results obtained, are presented. So far, our experience suggests that the multiple binding-site model is superior to all others tested, both in covering a wider range of assay types and in respect of the other advantages outlined above.

REFERENCES

1. Ekins, RP and Newman, GB (1970) *Acta Endocr. (Kbh.) Suppl. 147,* 11
2. Ekins, RP, Newman, GB and O'Riordan, JLH (1970) In *Statistics in Endocrinology.* (Ed) JW McArthur and T Colton. MIT Press, Mass. Page 345
3. Rodbard, D and Cooper, JA (1970) In *In vitro Procedures with Radioisotopes in Medicine.* International Atomic Energy Agency, Vienna. Page 659
4. Hales, CH and Randle, PJ (1963) *Biochem. J. 88,* 137
5. Meinert, CL and McHugh, RB (1968) *Math. Biosci. 2,* 319
6. Burger, HG, Lea, VWK and Rennie, GC (1973) *J. Lab. Clin. Med. 80,* 302
7. Fernandez, AA and Loeb, HG (1975) *Clin. Chem. 21,* 1113
8. Taljedal, I-B and Wold, S (1970) *Biochem. J. 119,* 139
9. Weaver, CK and Cargille, CM (1971) *J. Lab. Clin. Med. 77,* 661
10. Healy, MJR (1972) *Biochem. J. 130,* 207
11. Rodbard, D and Hutt, SM (1974) In *Radioimmunoassay and Related Procedures in Medicine.* International Atomic Energy Agency, Vienna. Vol. 1, Page 165
12. Malan, PG, Newman, GB and Ekins, RP (1973) *Acta Endocr. (Kbh.) Suppl. 177,* 99
13. Marschner, I, Erhardt, F and Scriba, PC (1973) In *Proc. Symp. on RIA and Related Procedures in Clin. Med. & Research.* Istanbul, Turkey
14. Reinsch, CH (1971) *Numer. Math. 16,* 451
15. Cox, MG and Hayes, JG (1973) *Report NAC 26,* National Physical Laboratory, Teddington, Middlesex, U.K.
16. Malan, PG, Seagroatt, V, Cox, MG and Ekins, RP (in preparation)
17. Gill, PE, Murray, W, Picken, SM, Wright, MH and Long, EMR National Physical Laboratory Algorithms Library, Ref. No. E4/32/F; BCMNAF. NPL, Teddington, Middlesex, U.K.
18. Powell, DR and Macdonald, JR (1972)
15, 148

EVALUATION AND DIAGNOSTIC INTERPRETATION OF TYPICAL ELECTROPHORETIC DIAGRAMS WITH THE AID OF A COMPUTER

Ch Langfelder, BSc(Mech)Eng
Dr D Neumeier, Priv Doz
Dr A Fateh-Moghadam
Prof Dr M Knedel
Institut für Klinische Chemie am
Klinikum Grosshadern der Universität München,
Postfach 701 260, D-8000 München 70

INTRODUCTION

The interpretation of protein concentration values and their relative displacement is of primary importance in the diagnosis of a number of diseases. For this reason serum electrophoresis is part of the laboratory routine in clinical chemistry.

Investigations by Mario Werner (1) in 1973 have shown that the information content and consequently the diagnostic usefulness gained from electrophoretic separation is only about 10 per cent less than (and hence in practice equal to) the information obtained by the quantitative determination of 14 single proteins in serum. In particular the electrophoretic report is the basis for frequently unexpected tentative diagnoses of paraproteinaemia. The information content in this case is superior to that gained by the quantitative determination of the immunoglobulins.

The number of electrophoretic reports required daily in a clinic is fairly high and continues to rise. The necessity to automate the method could not be overlooked despite some inherent difficulties in the various processing stages. With the installation of a computer three further objectives were to be reached:

1. A standardised form of identification and quantitative determination of the individual fractions.
2. An interpretation of the change in the relative concentration of the fractions.
3. A differentiation method for an increase in monoclonal and polyclonal gamma globulins.

METHOD

The electrophoretic separation of seven samples at a time on cellulose acetate film was carried out under the usual conditions. The analog values were determined with a Zeiss KFE 4 photometer. This instrument was connected to an 11/40 PDP computer equipped with a rack mounted laboratory subsystem. Because a number of variables cause differing conditions from one run to the next a pooled standard serum with two or more known peaks is included each time, a method of external standardisation common in gas chromatography practice. The positions of one or two fractions, e.g. α_1 or α_2 globulin, relative to a third fraction, e.g. albumin, permit the reproducible identification of the other fractions in all pherogrammes in which at least two of these so called leading fractions are recognised, as their positions can be established within extremely narrow limits. Thus a comparison of all apparent electrophoretic mobilities is possible.

A spline technique with a variable smoothing factor is used to smooth the curves output by the A-D converter (2). The identification and subsequent quantitative determination of the fractions are carried out by means of a simple window technique. The fractions are grouped on the basis of their normalised migration distances, obtained from the migration times of the leading fractions in the pooled standard serum. Mid-window values and standard deviations are stored in a parameter file, which also contains the parameters required for the curve presentation and text positions. Individual pherogram can therefore be reanalysed or plotted in another format by merely changing the necessary parameters. The laboratory report used in our institute is shown in Figure 1.

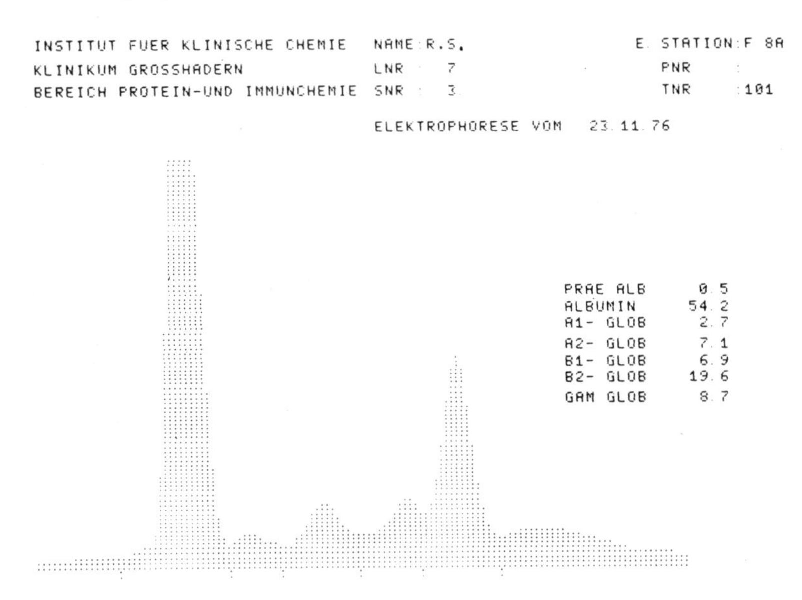

Figure 1. Electrophoresis laboratory report

STATISTICAL ANALYSES

The interassay precision was determined under routine laboratory conditions for the individual fractions of a pooled serum and found to be similar to values obtained for a manual evaluation.

The repeatability of a calculation was tested on a pherogram showing an antibody deficiency syndrome. The values obtained for the mean and coefficient of variation for the γ-globulin fraction were 3.31 per cent and 2.86 per cent respectively (Figure 2).

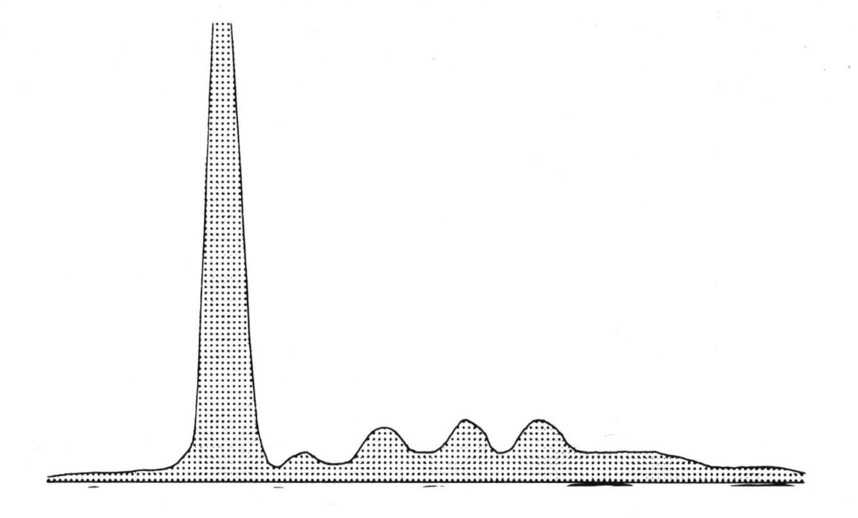

Pat: HA antibody deficiency syndrome

	praealbumin	*albumin*	*α_1-globulin*
MV	0.48	76.33	3.71
SD	0.06	0.22	0.03
VC	12.51	0.29	0.83
	α_2-globulin	*β-globulin*	*γ-globulin*
MV	6.66	9.20	3.31
SD	0.08	0.07	0.09
VC	1.15	0.71	2.86

Figure 2. Repeatability of calculation (FFE4/EDV), n = 20

To test the reproducibility of peak identification a serum with a paraprotein in the β/γ range was evaluated on six different days using a different position on the acetate foil each time. The results are shown in Figure 3. Using the normalised migration distance the paraprotein peak was ordered each time to the β_2 fraction.

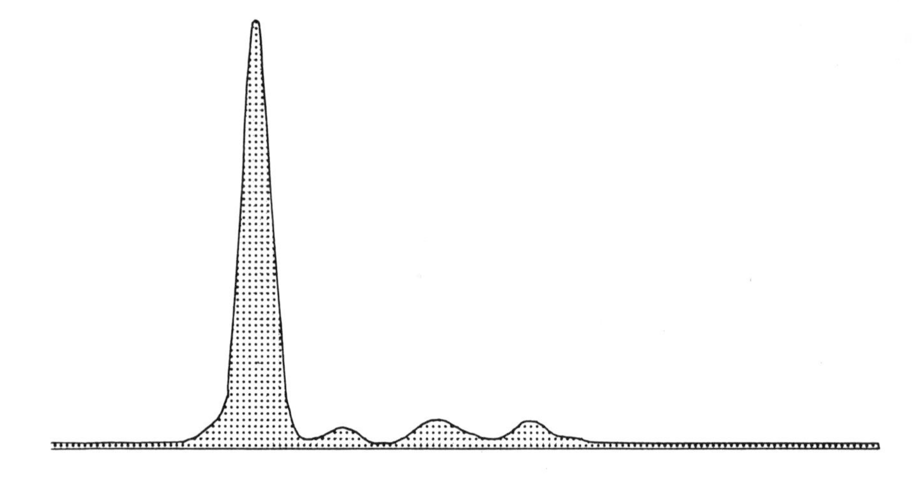

Differing sample sequence Pat: HJ paraprotein

	praealbumin	*albumin*	α_1*-globulin*
MV	0.47	62.37	2.77
SD	2.19	1.88	0.28
VC	88.83	3.01	10.14

	α_2*-globulin*	β*-globulin*	γ_1*-globulin*	γ_2*-globulin*
MV	8.37	8.87	9.76	7.22
SD	0.34	0.70	0.50	1.18
VC	4.12	7.90	5.10	16.40

Figure 3. Interassay variance for peak identification, n = 6

DIAGNOSTIC INTERPRETATION OF PHEROGRAMS

In the case of acute and chronic infections, nephrotic syndromes, hyperglobulinaemia, hypogammaglobulinaemia and dysproteinaemia the grouping of a fraction on the basis of concentration value and relative mobility is readily carried out with the aid of a computer as shown by

Nennstiel (3). The differentiation of monoclonal and polyclonal γ-globulin increases is not so easily performed.

Sera exhibiting monoclonal and polyclonal gamma globulin increases from 78 and 40 patients respectively were evaluated by the computer. In addition to the usual output, the peak geometry of the individual fractions was determined according to the characteristics shown in Figure 4. The trapezium that resulted from the longest linear section on the

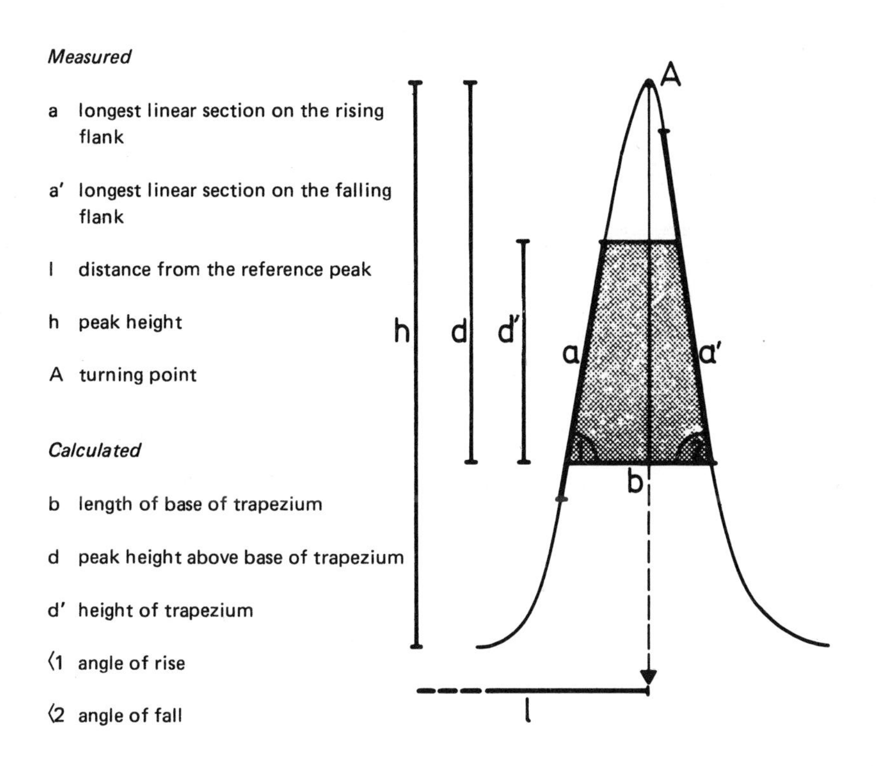

Figure 4. Peak geometry

rising and falling flanks of the curve provides the basis for nearly all subsequent decision variables. These are shown in Figure 5 and include the trapezium height, the trapezium area, the trapezium area/peak ratio, the angles of rise and fall, the trapezium base/peak height and trapezium base/peak area ratios. Two further variables were defined as the peak area to the albumin and β fraction areas. The absolute migration distance and deviation from the reference migration distance for a particular peak conclude the variable list. The shape and size of the α_2 fraction from 20 normal patient sera provided the reference values for the above variables.

1. height of trapezium
2. area of trapezium
3. area of trapezium/area of peak
4. angle of rise
5. angle of fall
6. width of trapezium/peak height
7. width of trapezium/peak area
8. peak area/peak$_{albumin}$
9. peak area/peak area$_{beta}$
10. migration time
11. deviation from reference migration time

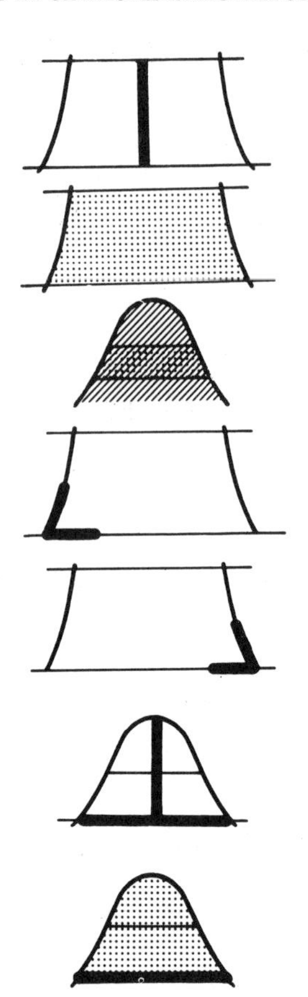

Figure 5. Decision variables

Prior to calculating a difference function of the form

$$\sum_{i=1}^{9} x_i f_i = x_1 f_1 + x_2 f_2 + \ldots x_9 f_9$$

(where x represents a measured variable such as the trapezium height and f is a weighting factor derived by methods of discriminance analysis) a discriminance analysis using the peak geometry parameters was made taking two groups of data at a time. The resultant discriminant functions and the grouping are shown in Figure 6.

The best discriminant function was obtained for the differentiation of normal and polyclonal gammopathy, while separation was poorest in the case of normal and monoclonal gammopathy, with five test data being grouped inconclusively. A visual control of these curves would have been equally difficult as these gammopathies exhibited an extremely small M-gradient.

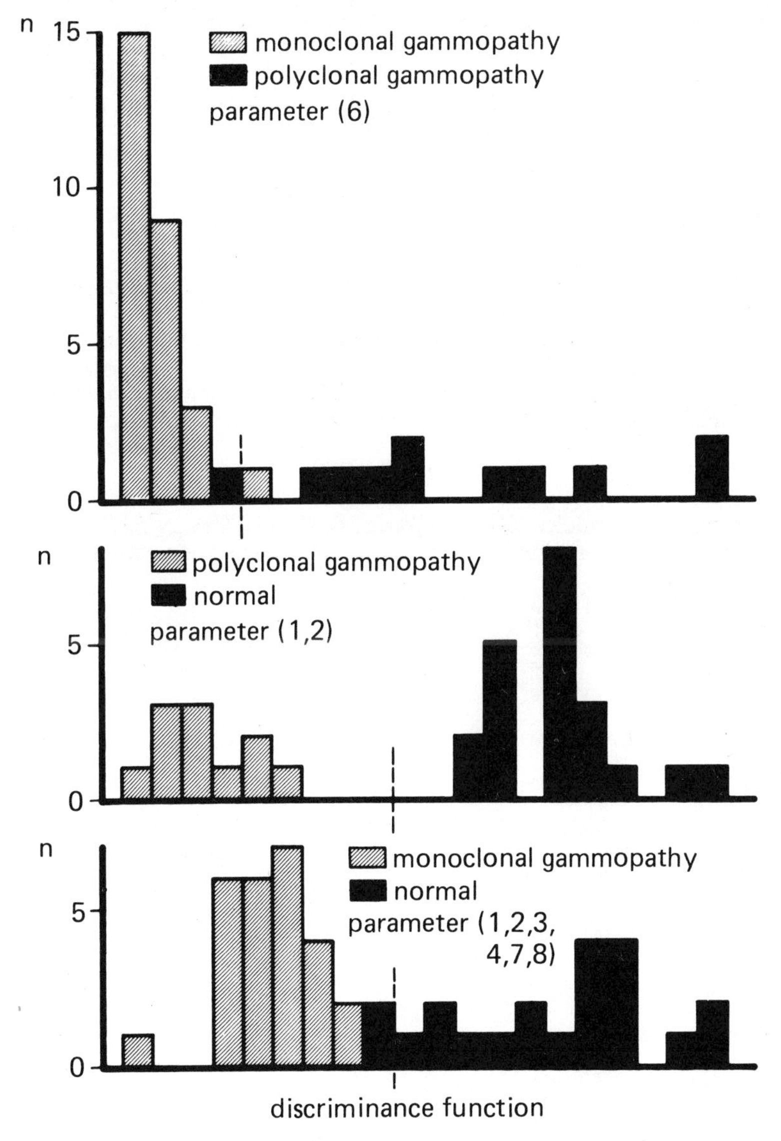

Figure 6. Discriminance function

The latter discriminant function was entered into the parameter file and further sera exhibiting monoclonal gamma globulin increases were evaluated by the computer. Of 53 sera, 46 were grouped correctly while 7 produced an error of type I (false negative). By shifting the discriminance function value, three errors of type II (false positive) were introduced while two errors of type I remained. Using the original discriminance function value and the migration distance variable improves the result to 51 correct groupings and 2 errors of type I. An inspection of the curves incorrectly grouped shows that curve form (largely parallel to the time axis) before and after the monoclonal peak might be included as a further parameter in the discriminance analysis to improve the separation function.

CONCLUSIONS

The method described above leads to a considerable gain in time due to the use of large acetate films.

Acceptable inter- and intravariance results make a comparison with conventional evaluation methods possible.

The use of a reference serum permits the reproducible identification of individual protein functions.

The use of a computer allows the selection of a group of parameters that matches a tentative diagnosis with an electrophoretic diagram.

REFERENCES

1. Werner, M, Brooks, SH and Wette, R (1973) *Human Pathol. 4,* 17
2. Späth, H (1973) Spline-Algorithmen zur Konstruktion glatter Kurven und Flächen
3. Nennstiel, H-S (1976) *Ärztl. Lab. 22,* 403

Index